T0271144

# Topics in Finite and Discrete Mathematics

Written for a broad audience of students in mathematics, computer science, operations research, statistics, and engineering, this textbook presents a short, lively survey of several fascinating noncalculus topics in modern applied mathematics. Coverage includes probability, mathematical finance, graphs, linear programming, statistics, computer science algorithms, and groups. A key feature is the abundance of interesting examples not normally found in standard finite mathematics courses, such as options pricing and arbitrage, tournaments, and counting formulas.

The author assumes a level of mathematical sophistication at the beginning calculus level; that is, students should have had at least a course in precalculus, and the added sophistication attained from studying calculus would be useful.

Sheldon M. Ross is a professor in the Department of Industrial Engineering and Operations Research at the University of California at Berkeley. He received his Ph.D. in statistics at Stanford University in 1968 and has been at Berkeley ever since. He has published nearly 100 articles as well as a variety of textbooks in the areas of statistics and applied probability. He is the founding and continuing editor of the journal *Probability in the Engineering and Informational Sciences,* a fellow of the Institute of Mathematical Statistics, and a recipient of the Humboldt U.S. Senior Scientist Award.

# Topics in Finite and Discrete Mathematics

SHELDON M. ROSS
*University of California at Berkeley*

CAMBRIDGE
UNIVERSITY PRESS

# CAMBRIDGE
## UNIVERSITY PRESS

32 Avenue of the Americas, New York NY 10013-2473, USA

Cambridge University Press is part of the University of Cambridge.

It furthers the University's mission by disseminating knowledge in the pursuit of
education, learning and research at the highest international levels of excellence.

www.cambridge.org
Information on this title: www.cambridge.org/9780521772594

First published 2000

*A catalogue record for this publication is available from the British Library*

*Library of Congress Cataloguing in Publication data*
Ross, Sheldon M.
Topics in finite and discrete mathematics / Sheldon M. Ross.
p. cm.
ISBN 0-521-77259-1 (hb) – ISBN 0-521-77571-X (pbk.)
I. Mathematics. I. Title.
QA39.2.R65485 2000
510- dc21 99–054713

ISBN 978-0-521-77259-4 Hardback
ISBN 978-0-521-77571-7 Paperback

*To* Rebecca

# Contents

# Preface

This text surveys many of the topics taught in discrete and finite mathematics courses. The topics chosen are widely applied in present-day industry and, at the same time, are mathematically elegant. Chapter 1 begins with such preliminaries as sets, mathematical induction, functions, and the division algorithm of algebra. Chapters 2 and 3 present combinatorics and probability. Chapter 4 introduces the modern approach to finance; it presents the concept of arbitrage and the arbitrage theorem and then uses them to analyze the no-arbitrage costs of options. Chapters 5 and 6 deal with graphs and their many applications. Chapter 7 introduces linear programming. Among other applications, we use the duality theorem to derive the arbitrage theorem as well as the minimax theorem of game theory. Chapter 8 presents sorting and searching techniques that are useful in computer science. Chapter 9 introduces the subject matter of statistics, presenting both its descriptive and inferential side. Chapter 10 deals with groups and permutations.

This book can be used for a course in discrete mathematics, or for one in finite mathematics, or for any course dealing with non–calculus-based applied mathematics. Calculus itself is not required, and a pre-calculus course should suffice as a prerequisite; the added mathematical sophistication attained from studying calculus would be useful. The text evolved from a seminar designed to introduce first-year undergraduates with a strong quantitative bent to the possibilities inherent in mathematics. Consequently, a key feature of the course, as well as of the text, is the emphasis on interesting examples.

# 1. Preliminaries

## 1.1 Sets

A *set* is a collection of elements. If the set $A$ consists of the $n$ elements $a_1, a_2, \ldots, a_n$ then we express this by writing

$$A = \{a_1, a_2, \ldots, a_n\}.$$

Thus, for instance, the set consisting of all the integers between 6 and 10 is given by

$$B = \{6, 7, 8, 9, 10\}.$$

A set can be defined either by specifying all its elements, as just shown, or by specifying a defining property for its elements. Thus, the set $B$ could have been defined as

$$B = \{\text{integers } i : |8 - i| \leq 2\}.$$

That is, $B$ could have been defined as the set of all integers $i$ such that the distance between $i$ and 8 is less than or equal to 2.

A set consisting of a finite number of elements is said to be a *finite set*, whereas one consisting of an infinite number of elements is said to be an *infinite set*. The set $\mathcal{N}$ of all the nonnegative integers is an example of an infinite set. It is convenient to define the set that does not consist of any elements; we call this the *null* set and denote it by $\emptyset$.

We use the notation $a \in A$ to indicate that $a$ is an element of $A$, and we use the notation $a \notin A$ to indicate that $a$ is not a member of $A$.

**Example 1.1a** Let $S$ be the set of all possible outcomes when a pair of dice are rolled. By an "outcome" we mean the pair $(i, j)$, where $i$ is the number of the side on which the first die lands and $j$ is the number of the side for the second die. Then, the set of all outcomes that result in the sum of the dice being equal to 7 can be expressed as

$$S_7 = \{(1, 6), (2, 5), (3, 4), (4, 3), (5, 2), (6, 1)\}$$

or, alternatively, as

$$S_7 = \{(i, j) \in S : i + j = 7\}. \qquad \square$$

If every element of $A$ is also an element of $B$, then we say that $A$ is a *subset* of $B$ and write $A \subset B$. By this definition, every set is a subset of itself and hence, for example, $A \subset A$. Also, since there are no elements in the null set, it follows that every element of $\emptyset$ is also an element of $A$; thus, $\emptyset$ is a subset of every other set. If $A \subset B$ and $B \subset A$ then we may write $A = B$. That is, the sets $A$ and $B$ are said to be equal if every element of $A$ is in $B$ and every element of $B$ is in $A$.

If $A$ and $B$ are sets then we define the new set $A \cup B$, called the *union* of $A$ and $B$, to consist of all elements that are in $A$ or in $B$ (or in both). Also, we define the *intersection* of $A$ and $B$, written either as $A \cap B$ or just $AB$, to consist of all elements that are in both $A$ and $B$.

**Example 1.1b**    In Example 1.1a, if $A$ is the set of all outcomes for which the sum of the dice is 5 and if $B$ is the set of outcomes for which the value of the second die exceeds that of the first die by the amount 3, then we have

$$A = \{(1, 4), (2, 3), (3, 2), (4, 1)\} \quad \text{and} \quad B = \{(1, 4), (2, 5), (3, 6)\};$$

also,

$$A \cup B = \{(1, 4), (2, 3), (3, 2), (4, 1), (2, 5), (3, 6)\} \quad \text{and}$$

$$AB = \{(1, 4)\}.$$

If we define $C$ to be the set of all outcomes whose sum is equal to 6, then $AC = \emptyset$ because there are no outcomes whose sum is both 5 and 6.    $\square$

A set is said to be a *universal set* if it contains (as subsets) all other sets under consideration. Let $\mathcal{U}$ be a universal set. For any set $A$, the set $A^c$, called the *complement* of $A$, is defined to be the set containing all the elements of the universal set $\mathcal{U}$ that are *not* in $A$.

*Venn diagrams* are often used to graphically represent sets. The universal set $\mathcal{U}$ is represented as consisting of all the points in a large rectangle, and sets are represented as consisting of all the points in circles within the rectangle. Particular sets of interest are indicated by shading

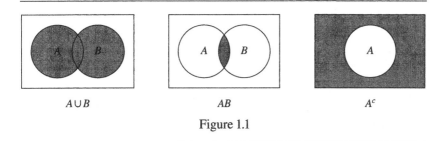

$A \cup B$        $AB$        $A^c$

Figure 1.1

appropriate regions of the diagram. For instance, the Venn diagrams of Figure 1.1 indicate the sets $A \cup B$, $AB$, and $A^c$.

The operation of forming unions and intersections of sets obey certain rules that are similar to the rules of algebra. We list a few of them as follows:

Commutative laws: $A \cup B = B \cup A$, $AB = BA$.
Associative laws: $(A \cup B) \cup C = A \cup (B \cup C)$, $(AB)C = A(BC)$.
Distributive laws: $(A \cup B)C = AC \cup BC$, $AB \cup C = (A \cup C)(B \cup C)$.

These relations are verified by showing that any element that is contained in the left-hand set is also contained in the right-hand one, and vice versa. For instance, to prove that

$$(A \cup B)C = AC \cup BC,$$

note that if $x \in (A \cup B)C$ then $x \in C$ and $x$ is also in either $A$ or $B$. If $x \in A$, then it is in $AC$ and so is in $AC \cup BC$; similarly, if $x \in B$, then it is in $BC$ and so is in $AC \cup BC$. Thus, $x \in AC \cup BC$, showing that

$$(A \cup B)C \subset AC \cup BC.$$

To go the other way, suppose that $y \in AC \cup BC$. Then $y$ is either in both $A$ and $C$ or in both $B$ and $C$. Therefore, we can conclude that $y$ is in $C$ and is in at least one of the sets $A$ and $B$. But this means that $y \in (A \cup B)C$, showing that

$$AC \cup BC \subset (A \cup B)C,$$

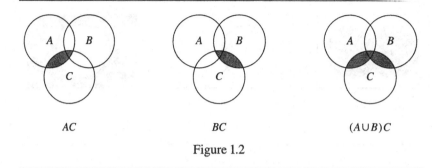

AC                    BC                    $(A \cup B)C$

Figure 1.2

and the verification is complete. (The result could also be shown by using Venn diagrams; see Figure 1.2.)

We also define the intersection and union of more than two sets. Specifically, for sets $A_1, \ldots, A_n$ we define $\bigcup_{i=1}^{n} A_i$, the union of these sets, to consist of all elements that are in $A_1$, or in $A_2$, or in $A_3$, $\ldots$, or in $A_n$; that is, $\bigcup_{i=1}^{n} A_i$ is the set of all elements that are in at least one of the sets $A_i$, $i = 1, \ldots, n$. Similarly, we define $\bigcap_{i=1}^{n} A_i$, the intersection of these sets, to consist of all elements that are in each of the sets $A_i$, $i = 1, \ldots, n$.

## 1.2    Summation

If we let $s$ be the sum of the four numbers $x_1, x_2, x_3, x_4$ then we can write

$$s = x_1 + x_2 + x_3 + x_4.$$

More compactly, we can use the summation notation $\sum$. Using this latter notation, we write

$$s = \sum_{i=1}^{4} x_i,$$

which means that $s$ is equal to the sum of the $x_i$ values as $i$ ranges from 1 to 4. More generally, for $j \leq n$, we use the notation

$$s = \sum_{i=j}^{n} x_i$$

to mean that

$$s = x_j + x_{j+1} + \cdots + x_n.$$

**Example 1.2a**   If $x_i = i^2$, find $\sum_{i=3}^{6} x_i$.

*Solution.*

$$\sum_{i=3}^{6} x_i = x_3 + x_4 + x_5 + x_6 = 9 + 16 + 25 + 36 = 86. \qquad \square$$

If $S$ is a specified set of integers, then we use the notation

$$\sum_{i \in S} x_i$$

to represent the sum of all the values $x_i$ that have indices in $S$.

**Example 1.2b**   If $S = \{2, 4, 6\}$ then

$$\sum_{i \in S} x_i = x_2 + x_4 + x_6. \qquad \square$$

Consider the sum $T = \sum_{i=0}^{2} x_{2+i}$. Because $T$ is equal to $x_2 + x_3 + x_4$, it follows that we can also express $T$ as $T = \sum_{j=2}^{4} x_j$. Therefore, we see that

$$\sum_{i=0}^{2} x_{2+i} = \sum_{j=2}^{4} x_j.$$

When equating the right-hand summation to the left, we say that we are making the change of variable $j = 2 + i$. That is, summing the values $x_{2+i}$ as $i$ ranges from 0 to 2 is the same as summing the values $x_j$ as $j$ ranges from 2 to 4.

**Example 1.2c**   Making the change of variable $j = n - i$ in the summation $\sum_{i=0}^{n} x_{n-i}$ gives the equivalent sum $\sum x_j$ as $j$ ranges between $n$ and 0. That is,

$$\sum_{i=0}^{n} x_{n-i} = \sum_{j=0}^{n} x_j. \qquad \square$$

We are sometimes interested in numbers that are expressed in the form $x_{i,j}$, where $i$ and $j$ both take values in some region. A quantity that is often of interest is the following "double sum" $D = \sum_{i=1}^{n} \sum_{j=1}^{m} x_{i,j}$, where

$$\sum_{i=1}^{n} \sum_{j=1}^{m} x_{i,j} = \sum_{i=1}^{n} \left( \sum_{j=1}^{m} x_{i,j} \right).$$

Now arrange the numbers $x_{i,j}$ ($i = 1, \ldots, n$, $j = 1, \ldots, m$) in the following row–column array, which has the number $x_{i,j}$ in row $i$, column $j$.

| $x_{1,1}$ | $x_{1,2}$ | $x_{1,3}$ | $\cdots$ | $x_{1,j}$ | $\cdots$ | $x_{1,m}$ |
|---|---|---|---|---|---|---|
| $x_{2,1}$ | $x_{2,2}$ | $x_{2,3}$ | $\cdots$ | $x_{2,j}$ | $\cdots$ | $x_{2,m}$ |
| $\vdots$ | $\vdots$ | $\vdots$ | $\vdots$ | $\vdots$ | $\vdots$ | $\vdots$ |
| $x_{i,1}$ | $x_{i,2}$ | $x_{i,3}$ | $\cdots$ | $x_{i,j}$ | $\cdots$ | $x_{i,m}$ |
| $\vdots$ | $\vdots$ | $\vdots$ | $\vdots$ | $\vdots$ | $\vdots$ | $\vdots$ |
| $x_{n,1}$ | $x_{n,2}$ | $x_{n,3}$ | $\cdots$ | $x_{n,j}$ | $\cdots$ | $x_{n,m}$ |

Because $\sum_{j=1}^{m} x_{i,j}$ is just the sum of the $m$ array elements in row $i$, it follows that the double sum $D$ is equal to the sum of the row sums. In other words, $D$ is equal to the sum of all the elements in the array. Since the sum of all the array values can also be obtained by adding all the column sums and since the sum of the values of column $j$ is $\sum_{i=1}^{n} x_{i,j}$, we have the following result.

**Proposition 1.2.1**

$$\sum_{i=1}^{n} \sum_{j=1}^{m} x_{i,j} = \sum_{j=1}^{m} \sum_{i=1}^{n} x_{i,j}.$$

A corollary of this proposition is the following useful result.

**Corollary 1.2.1**

$$\sum_{i=1}^{n} \sum_{j=1}^{i} x_{i,j} = \sum_{j=1}^{n} \sum_{i=j}^{n} x_{i,j}.$$

*Proof.* Consider data values $x_{i,j}$, where $i$ and $j$ both takes values from 1 to $n$ and where $x_{i,j} = 0$ when $j > i$. Then apply Proposition 1.2.1.    $\square$

A pictorial proof of Corollary 1.2.1 is obtained by noting that its left-hand side,

$$\sum_{i=1}^{n}\sum_{j=1}^{i} x_{i,j} = \sum_{j=1}^{1} x_{1,j} + \sum_{j=1}^{2} x_{2,j} + \cdots + \sum_{j=1}^{n} x_{n,j},$$

is equal to the sum of all the row sums whereas the right-hand side,

$$\sum_{j=1}^{n}\sum_{i=j}^{n} x_{i,j} = \sum_{i=1}^{n} x_{i,1} + \sum_{i=2}^{n} x_{i,2} + \cdots + \sum_{i=n}^{n} x_{i,n},$$

is the sum of all the column sums in the following array.

$$
\begin{array}{ccccc}
x_{1,1} \\
x_{2,1} & x_{2,2} \\
x_{3,1} & x_{3,2} & x_{3,3} \\
\vdots & \vdots & \vdots & \ddots \\
x_{n,1} & x_{n,2} & x_{n,3} & \cdots & x_{n,n}
\end{array}
$$

**Example 1.2d**

$$\sum_{i=1}^{3}\sum_{j=1}^{i}(i-j) = \sum_{j=1}^{1}(1-j) + \sum_{j=1}^{2}(2-j) + \sum_{j=1}^{3}(3-j)$$
$$= 0 + 1 + 3 = 4,$$

$$\sum_{j=1}^{3}\sum_{i=j}^{3}(i-j) = \sum_{i=1}^{3}(i-1) + \sum_{i=2}^{3}(i-2) + \sum_{i=3}^{3}(i-3)$$
$$= 3 + 1 + 0 = 4. \qquad \square$$

Since $x \sum_{j=1}^{m} y_j = \sum_{j=1}^{m} xy_j$, it follows that

$$\left(\sum_{i=1}^{n} x_i\right)\left(\sum_{j=1}^{m} y_j\right) = \sum_{j=1}^{m}\left(\sum_{i=1}^{n} x_i\right) y_j = \sum_{j=1}^{m}\sum_{i=1}^{n} x_i y_j.$$

**Example 1.2e**  Expand $(x_1 + \cdots + x_n)^2$.

**Solution.**

$$\left(\sum_{i=1}^{n} x_i\right)^2 = \left(\sum_{i=1}^{n} x_i\right)\left(\sum_{j=1}^{n} x_j\right)$$

$$= \sum_i \sum_j x_i x_j$$

$$= \sum_i \left(x_i x_i + \sum_{j \neq i} x_i x_j\right)$$

$$= \sum_i x_i^2 + \sum_i \sum_{j \neq i} x_i x_j.$$

For instance, the preceding yields

$$(x_1 + x_2)^2 = x_1^2 + x_2^2 + x_1 x_2 + x_2 x_1 = x_1^2 + x_2^2 + 2x_1 x_2. \qquad \square$$

Similar to the notation for summations is our notation $\prod$ for products,

$$\prod_{i=1}^{n} x_i = x_1 x_2 \cdots x_n.$$

**Example 1.2f**

$$\prod_{i=1}^{4} i = 1 \cdot 2 \cdot 3 \cdot 4 = 24. \qquad \square$$

## 1.3    Mathematical Induction

Suppose that we have an infinite collection of statements, denoted $S_1, S_2, \ldots$, and that we want to prove that they are all true. A proof by *mathematical induction* is obtained in the following manner:

(i)  first prove that $S_1$ is true;
(ii) then show that, for any $n$, whenever $S_n$ is true then $S_{n+1}$ is also true.

Once (i) and (ii) are established, then from (i) we know that $S_1$ is true; which implies by (ii) that $S_2$ is true; which implies that $S_3$ is true; and so on. Thus, it follows that all of the $S_n$ are true.

We now illustrate the use of mathematical induction by a series of examples.

**Example 1.3a**   Prove that there are $2^n$ subsets of a set consisting of $n$ elements.

*Solution.* In order to prove this by mathematical induction, we must first prove it for $n = 1$. But this is immediate, for if the set consists of a single element (i.e., if the set is $\{s\}$) then it has the two subsets $\emptyset$ and $\{s\}$, where $\emptyset$ is the empty set. Thus, part (i) of the mathematical induction approach is shown. To show part (ii), *assume* that the result is true for all sets of size $n$ (this is called the *induction hypothesis*) and then consider a set $S$ of size $n + 1$. Focus attention on one of the elements of $S$, call it $s$, and let $S'$ denote the set consisting of the $n$ other elements of $S$. Because every subset of $S$ that does not contain $s$ is a subset of $S'$, it follows from the induction hypothesis that there are $2^n$ subsets of $S$ that do not contain $s$. On the other hand, since any subset of $S$ that contains $s$ can be obtained by adding $s$ to a subset of $S'$, it also follows from the induction hypothesis that there are $2^n$ of these subsets. Thus the total number of subsets of $S$ is

$$2^n + 2^n = 2^n(1+1) = 2^{n+1},$$

and the result is proved.                            □

**Example 1.3b**   For integer $n$, which is larger: $2^n$ or $n^2$?

*Solution.* Let us try a few cases:

$$2^1 = 2, \qquad 1^2 = 1;$$
$$2^2 = 4, \qquad 2^2 = 4;$$
$$2^3 = 8, \qquad 3^2 = 9;$$
$$2^4 = 16, \qquad 4^2 = 16;$$
$$2^5 = 32, \qquad 5^2 = 25;$$
$$2^6 = 64, \qquad 6^2 = 36.$$

Thus, based on this enumeration, a reasonable conjecture is that $2^n > n^2$ for all values of $n \geq 5$. To prove this, we start by showing it to be true when $n = 5$; this was demonstrated by our preceding calculations. So now assume that, for some $n$ ($n \geq 5$),

$$2^n > n^2.$$

We must show that the preceding implies that $2^{n+1} > (n + 1)^2$, which may be accomplished as follows.

First, note that

$$2^{n+1} = 2 \cdot 2^n > 2n^2,$$

where the inequality follows from the induction hypothesis. Hence, it will suffice to show that, for $n \geq 5$,

$$2n^2 \geq (n + 1)^2$$

or (equivalently)

$$2n^2 \geq n^2 + 2n + 1$$

or

$$n^2 - 2n - 1 \geq 0$$

or

$$(n - 1)^2 - 2 \geq 0$$

or

$$n - 1 \geq \sqrt{2},$$

which follows because $n \geq 5$.                                    □

**Example 1.3c**   Derive a simple expression for the following function:

$$f(n) = \frac{1}{1 \cdot 2} + \frac{1}{2 \cdot 3} + \frac{1}{3 \cdot 4} + \cdots + \frac{1}{n(n + 1)}.$$

*Solution.*  Again, let us begin by calculating the value of $f(n)$ for small values of $n$, hoping to discover a general pattern that we can then prove by mathematical induction.  Such a calculation gives

$$f(1) = 1/2,$$
$$f(2) = 1/2 + 1/6 = 2/3,$$
$$f(3) = 2/3 + 1/12 = 3/4,$$
$$f(4) = 3/4 + 1/20 = 4/5.$$

Thus, a reasonable conjecture is that

$$f(n) = \frac{n}{n+1}.$$

Let us now prove this by induction. Since it is true when $n = 1$, assume that it is valid also for some other $n$ and consider $f(n+1)$. We have

$$f(n+1) = \frac{1}{1 \cdot 2} + \frac{1}{2 \cdot 3} + \cdots + \frac{1}{n(n+1)} + \frac{1}{(n+1)(n+2)}$$

$$= f(n) + \frac{1}{(n+1)(n+2)}$$

$$= \frac{n}{n+1} + \frac{1}{(n+1)(n+2)} \quad \text{(by the induction hypothesis)}$$

$$= \frac{n(n+2) + 1}{(n+1)(n+2)}$$

$$= \frac{(n+1)^2}{(n+1)(n+2)}$$

$$= \frac{n+1}{n+2}.$$

Thus, the result is established. (As in any situation where one has proven a particulary nice result by mathematical induction, it pays to see if there is a more direct argument that establishes and also *explains* the result; see Exercise 1.18.)    □

**Example 1.3d**  If one has unlimited access to five-cent and seven-cent stamps, show that any postage value greater than or equal to 24 cents can be exactly met.

*Solution.*  First note that a postage of 24 can be obtained by 2 fives and 2 sevens. Now assume that for some $n \geq 24$ the postage value $n$ can be exactly hit with a combination of five- and seven-cent stamps, and suppose that we desire postage of value $n + 1$. To obtain this exact amount, consider the combination that adds up to $n$. If it contains at least 2 seven-cent stamps, then trade 2 sevens for 3 fives to obtain the postage value $n + 1$. If the combination adding to $n$ contains at least 4 fives, replace them by 3 sevens to obtain the value $n + 1$. Thus, the result is shown if the combination adding up to $n$ contains either at least 2 sevens or at least 4 fives. The alternative is that it contains at most 1 seven and at most 3 fives; but this would imply that $n \leq 22$, which is not the case. Thus, the result is shown.  □

**Example 1.3e**  Show that, for any positive integer $n$,

$$\sum_{i=1}^{n} i = \frac{n(n+1)}{2}.$$

*Solution.*  We need to show that

$$1 + 2 + \cdots + n = \frac{n(n+1)}{2}.$$

This is true for $n = 1$, since both sides are equal to 1. So let us assume that it is true for some integer $n$. To verify it for $n + 1$, we reason as follows:

$$1 + 2 + \cdots + n + n + 1$$
$$= \frac{n(n+1)}{2} + n + 1 \quad \text{(by the induction hypothesis)}$$
$$= (n+1)\left(\frac{n}{2} + 1\right)$$
$$= \frac{(n+1)(n+2)}{2},$$

and the induction proof is complete.  □

**Example 1.3f**  Verify that, for any value $x \neq 1$ and positive integer $n$,

$$\sum_{i=0}^{n} x^i = \frac{1 - x^{n+1}}{1 - x}.$$

*Solution.*  Let us use induction. When $n = 1$, the identity says that

$$1 + x = \frac{1 - x^2}{1 - x},$$

which is true because

$$1 - x^2 = (1 - x)(1 + x).$$

So assume that the identity is true for a specified $n$. To prove that it remains true when $n$ is increased by 1, note the following:

$$\sum_{i=0}^{n+1} x^i = \sum_{i=0}^{n} x^i + x^{n+1}$$

$$= \frac{1 - x^{n+1}}{1 - x} + x^{n+1} \quad \text{(by the induction hypothesis)}$$

$$= \frac{1 - x^{n+1} + x^{n+1} - x^{n+2}}{1 - x}$$

$$= \frac{1 - x^{n+2}}{1 - x}.$$

Thus, the identity is also valid for $n + 1$, which shows that it is true for all $n$. □

**Example 1.3g** In a round-robin tennis tournament, every pair of competitors play a match. Show that if such a tournament were played with $n$ players then there is a labeling of the players $p_1, p_2, \ldots, p_n$ such that

$$p_1 \text{ beat } p_2, \ p_2 \text{ beat } p_3, \ \ldots, \ p_{n-1} \text{ beat } p_n. \tag{1.1}$$

*Solution.* The verification is by induction. The result is immediate when $n = 2$, so suppose it to be true whenever there are $n$ players and consider the case when there are $n + 1$. Put one of the players, call her $p$, aside. Then, by the induction hypothesis, there is an ordering of the other $n$ players such that (1.1) holds. If $p$ did not beat any of the other $n$ players then

$$p_1 \text{ beat } p_2, \ p_2 \text{ beat } p_3, \ \ldots, \ p_{n-1} \text{ beat } p_n, \ p_n \text{ beat } p.$$

On the other hand, if $p$ won at least one match then, with $i$ equal to the smallest integer such that $p$ beat $p_i$,

$$p_1 \text{ beat } p_2, \ \ldots, \ p_{i-1} \text{ beat } p, \ p \text{ beat } p_i, \ \ldots, \ p_{n-1} \text{ beat } p_n.$$

Thus the result is true whenever there are $n + 1$ players, which completes the induction proof.    □

The following result, although intuitively obvious, is quite useful.

**Proposition 1.3.1** *Every finite nonempty set of numbers A has a smallest and a largest element.*

**Proof.** We shall show by induction that $A$ always has a smallest and a largest element whenever $A$ is a set of $n$ numbers. This is true when $n = 1$ (since the lone number in $A$ is both the smallest and largest number of $A$), so assume it to be true for all sets of $n$ numbers. Let $A$ be a set consisting of $n + 1$ numbers, say $A = \{a_1, \ldots, a_n, a_{n+1}\}$. Then, by the induction hypothesis, the subset $\{a_1, \ldots, a_n\}$ has a smallest and largest element (say, $a_i$ and $a_j$ resp.). But then $A$ has a smallest element, namely the smaller of $a_i$ and $a_{n+1}$, and a largest element, namely the larger of $a_j$ and $a_{n+1}$. This completes the induction and, since a finite nonempty set must contain $n$ elements for some $n$, also establishes the result.    □

The well-ordering property of the integers is a simple consequence of Proposition 1.3.1.

**Corollary 1.3.1** (Well-Ordering Property of Positive Integers)  *Every set A containing at least one positive integer has a smallest positive integer.*

**Proof.** Let $n$ be a positive integer in $A$. Any integer in $A$ that is larger than $n$ cannot be the smallest positive integer in $A$. Hence it follows that, if the set $A_n = \{i : i \text{ is an integer}, i \in A, i \leq n\}$ has a smallest member, then that integer is also the smallest positive integer in $A$. But since $A_n$ is a finite set, it has a smallest member.    □

We now use mathematical induction to prove a well-known mathematical result.

**Proposition 1.3.2** (Hardy's Lemma)  *Consider two collections of numbers,*

$$a_1 \le a_2 \le \cdots \le a_n \quad and \quad b_1 \le b_2 \le \cdots \le b_n,$$

*and suppose that we have to make n disjoint pairs from these collections, each pair consisting of one a and one b. Then the sum of the products of the members of each pair is maximized when $a_i$ is paired with $b_i$ for each $i = 1, \ldots, n$.*

**Proof.** When $n = 2$ we must show that

$$a_1 b_1 + a_2 b_2 \ge a_1 b_2 + a_2 b_1,$$

which is equivalent to

$$a_2(b_2 - b_1) \ge a_1(b_2 - b_1)$$

or

$$(a_2 - a_1)(b_2 - b_1) \ge 0;$$

this is true beause both factors are nonnegative. So assume that the result is true whenever there are $n$ numbers in each collection, and suppose now that there are $n + 1$ values $a_1 \le \cdots \le a_{n+1}$ and $n + 1$ values $b_1 \le \cdots \le b_{n+1}$ to be paired up. Consider any pairing of the $n + 1$ a and b values in which $a_1$ is not paired with $b_1$ – rather, $a_1$ is paired with (say) $b_i$. Then, aside from this individual pairing, there remain $n$ members of each set to be paired up:

$$a_2, \ldots, a_i, a_{i+1}, \ldots, a_{n+1}$$

to be paired up with

$$b_1, \ldots, b_{i-1}, b_{i+1}, \ldots, b_{n+1}.$$

By the induction hypothesis, the pairing that maximizes the sum of the products from the remaining pairings – call this pairing $M$ – will pair $a_2$ (the smallest a) with $b_1$ (the smallest b). Thus the best pairing that pairs up $a_1$ with $b_i$ will also pair up $a_2$ with $b_1$. But by the result shown when $n = 2$, it is at least as good to pair up $a_1$ with $b_1$ and $a_2$ with $b_i$ and then pair the others as does $M$. Thus, we need only consider pairings that pair up $a_1$ and $b_1$; by the induction hypothesis, the best one of this type also pairs up $a_i$ with $b_i$ for each $i = 2, \ldots, n + 1$, which completes the proof. $\square$

The mathematical induction proof technique sometimes uses the following, "strong" version of induction.

**Strong Version of Mathematical Induction**   To prove that all the statements $S_1, S_2, \ldots$ are true:

(i)  prove that $S_1$ is true;
(ii) show that, for any $n$, if $S_1, \ldots, S_n$ are all true then $S_{n+1}$ is also true.

The strong version is valid because – once (i) and (ii) are established – from (i) we know that $S_1$ is true; which implies by (ii) that $S_2$ is true; which implies, since $S_1$ and $S_2$ are both true, that $S_3$ is true; and so on. Indeed, the strong version proof that all of the statements $S_n$ are true is equivalent to the standard mathematical induction proof of the statements $S_n^*$ ($n \geq 1$), where $S_n^*$ is the statement that $S_1, \ldots, S_n$ are all true.

**Example 1.3h**   Let $a_1 = 3$, $a_2 = 7$, and

$$a_n = 3a_{n-1} - 2a_{n-2}, \quad n = 3, 4, \ldots.$$

Find an explicit expression for $a_n$ and prove your result.

*Solution.* Let us start by evaluating some of the early values of $a_n$ in the hope of discovering a pattern. This yields

$$a_1 = 3,$$
$$a_2 = 7,$$
$$a_3 = 21 - 6 = 15,$$
$$a_4 = 45 - 14 = 31,$$
$$a_5 = 93 - 30 = 63,$$
$$a_6 = 189 - 62 = 127.$$

It is not difficult to spot that $a_n = 2^{n+1} - 1$ for all of the values of $n$ between 1 and 6. To prove that this holds for all $n$, assume that $a_k = 2^{k+1} - 1$ for all values less than or equal to $n$. Then

$$a_{n+1} = 3a_n - 2a_{n-1}$$
$$= 3(2^{n+1} - 1) - 2(2^n - 1)$$
$$= 3 \cdot 2^{n+1} - 3 - 2^{n+1} + 2$$
$$= 2 \cdot 2^{n+1} - 1,$$

which completes the induction proof since $2 \cdot 2^{n+1} = 2^{n+2}$.    □

## 1.4    Functions

A real-valued function is a rule that associates a real number to every element $x$ of a set $X$. The function is symbolically represented as $f$, and the value associated to the element $x$ is designated as $f(x)$. The set $X$ is called the *domain* of $f$.

**Example 1.4a**   If $X$ is the set of integers, then the function

$$f(i) = i^2$$

associates to each integer $i$ the value $i^2$.    □

**Definition**   Let $f$ be a function whose domain is the set of integers. We say that $f$ is an *increasing* function if, for every integer $i$,

$$f(i + 1) \geq f(i).$$

Similarly, we say that $f$ is a *decreasing* function if, for every integer $i$,

$$f(i + 1) \leq f(i).$$

**Example 1.4b**   Are the following functions increasing, decreasing, or neither?

(a)  $f(i) = 5i$.

(b)  $f(i) = i^2$.

(c)  $f(i) = \log(i)$.

(d)  $f(i) = \begin{cases} 0 & \text{if } i \text{ is even,} \\ 1 & \text{if } i \text{ is odd.} \end{cases}$

***Solution.*** The function in (a) is increasing. The function in (b) is increasing if the domain of the function is the set of nonnonegative integers; however, if the domain is the set of all integers then it is neither increasing nor decreasing. Assuming that the domain of the function in (c) is the set of positive integers, then the function is increasing. The function in (d) is neither increasing nor decreasing.    □

If $f$ is an increasing function on the integers, then it can be shown that

$$f(i) \leq f(j) \quad \text{if } i < j. \tag{1.2}$$

One way to establish (1.2) is to note the sequence of inequalities

$$f(i) \leq f(i+1) \leq f(i+2) \leq \cdots \leq f(j).$$

A more formal proof would be to use mathematical induction to prove that, for all $n \geq 0$,

$$f(n+i) \geq f(i).$$

The preceding is true when $n = 1$; assuming it true for $n$ yields

$$f(n+1+i) \geq f(n+i) \quad \text{(by the definition of an increasing function)}$$
$$\geq f(i) \quad \text{(by the induction hypothesis),}$$

which completes the more formal induction proof of equation (1.2).

If $f$ and $g$ are functions defined on the same domain $X$, then we say that

$$f \leq g \quad \text{(equivalently, } g \geq f\text{)}$$

if, for all $x \in X$,

$$f(x) \leq g(x).$$

Similarly, we say that

$$f = g$$

if, for all $x \in X$,

$$f(x) = g(x).$$

The function that associates the same value $c$ to every element in $X$ is said to be a *constant* function and is denoted by $c$. Hence, the notation

$$f \leq c$$

means that $f(x) \leq c$ for all $x \in X$. A function $f$ of the form

$$f(x) = a_0 + a_1 x + a_2 x^2 + \cdots + a_n x^n$$

is said to be a *polynomial* function. The next example uses mathematical induction to verify a sufficient condition for a polynomial function to be positive whenever $x \geq 1$.

**Example 1.4c**   Prove that

$$\sum_{i=0}^{n} a_i x^i > 0 \quad \text{for all } x \geq 1,$$

provided that

$$a_n > 0,$$
$$a_{n-1} + a_n > 0,$$
$$a_{n-2} + a_{n-1} + a_n > 0,$$
$$\vdots$$
$$a_0 + a_1 + a_2 + \cdots + a_{n-1} + a_n > 0.$$

*Solution.* Suppose the preceding conditions on $a_0, \ldots, a_n$ and assume that $x \geq 1$. Let

$$P(0) = a_n,$$
$$P(1) = a_{n-1} + x a_n = a_{n-1} + x P(0),$$
$$P(2) = a_{n-2} + x a_{n-1} + x^2 a_n = a_{n-2} + x P(1),$$
$$\vdots$$
$$P(j) = a_{n-j} + x a_{n-j+1} + \cdots + x^j a_n = a_{n-j} + x P(j-1),$$
$$j = 1, \ldots, n.$$

Thus, the objective is to show that

$$P(n) > 0 \quad \text{if } x \geq 1.$$

We will accomplish this by using mathematical induction to prove that, for all $j = 0, \ldots, n$,

$$P(j) \geq a_{n-j} + a_{n-j+1} + \cdots + a_n. \tag{1.3}$$

Since the RHS of (1.3) is assumed to be positive, the result would then be proven. Equation (1.3) holds when $j = 0$, so assume that

$$P(j) \geq a_{n-j} + a_{n-j+1} + \cdots + a_n > 0.$$

Then,

$$
\begin{aligned}
P(j + 1) &= a_{n-j-1} + xP(j) \\
&> a_{n-j-1} + P(j) \quad \text{(since } P(j) > 0 \text{ and } x \geq 1) \\
&\geq a_{n-j-1} + a_{n-j} + a_{n-j+1} + \cdots + a_n \\
&\qquad \text{(by the induction hypothesis),}
\end{aligned}
$$

and the proof by mathematical induction is complete.    $\square$

Functions on the same domain can be combined to form new functions. For instance, if $f$ and $g$ are functions on the integers then so are the functions $f + g$ and $fg$, defined by

$$f + g(i) = f(i) + g(i),$$
$$fg(i) = f(i)g(i).$$

That is, the values associated with $i$ by the functions $f + g$ and $fg$ are, respectively, $f(i) + g(i)$ and $f(i)g(i)$.

**Definition**    Let $f$ be a function whose domain is the set of integers, and define the function $g$ by

$$g(i) = f(i) - f(i - 1).$$

We say that $f$ is a *convex* function if $g$ is an increasing function; that is, $f$ is convex if for all $i$,

$$f(i + 1) - f(i) \geq f(i) - f(i - 1).$$

Similarly, we say that $f$ is a *concave* function if $g$ is a decreasing function; that is, if for all $i$,

$$f(i + 1) - f(i) \leq f(i) - f(i - 1).$$

**Example 1.4d**   Characterize the functions of Example 1.4b as to convexity and concavity.

*Solution.* (a) Since

$$f(i + 1) - f(i) = 5(i + 1) - 5i = 5,$$

it follows that $f$ is both convex and concave.
   (b) Since

$$f(i + 1) - f(i) = (i + 1)^2 - i^2 = 2i + 1,$$

it follows that $f$ is a convex function.
   (c) Note that

$$f(i + 1) - f(i) = \log(i + 1) - \log(i) = \log\left(\frac{i + 1}{i}\right) = \log\left(1 + \frac{1}{i}\right).$$

Since $\log(j)$ is an increasing function of $j$ and since $1 + 1/i$ is decreasing in $i$, it follows that $\log(i)$ is a concave function.
   (d) Since $f(i + 1) - f(i)$ is equal to 1 when $i$ is even and to $-1$ when $i$ is odd, it is neither an increasing nor a decreasing function. Hence, $f$ is neither convex nor concave.                                      □

Let $f$ be a function defined on the integers (i.e., its domain is the integers). For a vector of integers $\mathbf{x} = (x_1, \ldots, x_n)$, define the function $V(\mathbf{x})$ by

$$V(\mathbf{x}) = \sum_{j=1}^{n} f(x_j).$$

Let us say that $\mathbf{x}$ is a feasible vector if, for a specified integer $k$,

$$\sum_{j=1}^{n} x_j = nk.$$

Consider the problem of choosing a feasible vector to maximize $V(\mathbf{x})$.

**Proposition 1.4.1**   *If $f$ is concave, then $(k, k, \ldots, k)$ is a maximal vector. That is, for any feasible vector $\mathbf{x} = (x_1, \ldots, x_n)$,*

$$nf(k) \geq \sum_{j=1}^{n} f(x_j).$$

***Proof.*** Let $\mathbf{x} = (x_1, \ldots, x_n)$ be an arbitrary feasible vector for which not all $x_j$ are equal to $k$. Since $\sum_{j=1}^{n}(x_j/n) = k$, it follows that there will be indices $i$ and $j$ such that $x_i < k < x_j$. As a result, since $f$ is concave and $x_j \geq x_i + 1$, we have

$$f(x_j) - f(x_j - 1) \leq f(x_i + 1) - f(x_i)$$

or (equivalently)

$$f(x_i + 1) + f(x_j - 1) \geq f(x_i) + f(x_j),$$

which shows that the vector obtained from $\mathbf{x}$ by increasing $x_i$ by 1, decreasing $x_j$ by 1, and leaving the other components as is, is a feasible vector whose associated $V$ value is at least as large as $V(\mathbf{x})$. Repeating this procedure produces a sequence of feasible vectors, each with an associated $V$ value at least as large as the one preceding it. In addition, since each successive vector $\mathbf{y}$ has a strictly smaller value of $\sum_j |y_j - k|$, it follows that this sequence will eventually end. But since it can only end with the vector $k, k, \ldots, k$, we see that this vector is eventually reached; thus, its associated $V$ value is at least as large as $V(\mathbf{x})$. Since $\mathbf{x}$ is an arbitrary feasible vector, the result is proved.    $\square$

Proof of the following corollary is left as an exercise.

**Corollary 1.4.1**  *If $f$ is a convex function, then*

$$\min\left\{ \sum_{j=1}^{n} f(x_j) : \sum_{j=1}^{n} x_j = nk \right\} = nf(k).$$

For any function $f$ whose domain is the set of integers, let $\Delta f$ be the function defined by

$$\Delta f(i) = f(i) - f(i - 1).$$

Thus, "$f$ is increasing" is equivalent to

$$\Delta f \geq 0.$$

Also, let

$$\Delta^2 f = \Delta(\Delta f);$$

that is,

$$\Delta^2 f(i) = \Delta f(i) - \Delta f(i-1) = f(i) - f(i-1) - [f(i-1) - f(i-2)].$$

Thus, "$f$ is convex" is equivalent to

$$\Delta^2 f \geq 0.$$

## 1.5 The Division Algorithm

Let $n$ and $k$ be positive integers. Dividing $n$ by $k$ results in a quotient $q$ and remainder $r$, where $0 \leq r < k$ The formal statement of this is known as Euclid's division lemma.

**Proposition 1.5.1** (Euclid's Division Lemma)  *For any pair of positive integers $n$ and $k$, there are unique integers $q$ and $r$ where $0 \leq r < k$ and*

$$n = qk + r.$$

*Proof.* Let $\mathcal{N}$ denote the set of all nonnegative integers, and let

$$A = \{i \in \mathcal{N} : ik \leq n\}.$$

Since $A$ is clearly a finite nonempty set, it follows from Proposition 1.3.1 that it has a largest element. Call its largest element $q$, and let

$$r = n - qk \geq 0.$$

Now $r < k$, because if $r \geq k$ then

$$0 \leq r - k = n - (q+1)k,$$

which contradicts the fact that $q$ is the largest element in $A$. Hence, we have the representation

$$n = qk + r,$$

where $0 \leq r < k$. To prove uniqueness, suppose that there is a second representation

$$n = q_1 k + r_1,$$

where $0 \leq r_1 < k$. It follows that

$$(q - q_1)k = r_1 - r.$$

But unless $q = q_1$, the LHS of the preceding is an integral multiple of $k$ whereas the absolute value of the RHS is strictly less than $k$. Since this is impossible, we can conclude that $q = q_1$, which implies from the preceding equation that $r = r_1$. Hence, the representation is unique.    □

If $r = 0$ in the representation

$$n = qk + r, \quad 0 \le r < k,$$

then we say that $k$ is a *divisor* of $n$ and that $n$ is *divisible* by $k$.

If $n$ and $m$ are nonnegative integers, not both equal to 0, then we say that $d$ is their *greatest common divisor,* written as $d = \gcd(n, m)$, if:

 (i) $d$ is a divisor of both $n$ and $m$; and
 (ii) if $c$ is also a divisor of $n$ and $m$, then $c$ is also a divisor of $d$.

Thus, the greatest common divisor of $n$ and $m$ is a common divisor of $n$ and $m$ that is divisible by every other common divisor. We now show that the greatest common divisor always exists.

**Proposition 1.5.2**  *For nonnegative integers $n$ and $m$, not both equal to 0, $\gcd(n, m)$ exists.*

*Proof.* Let $\mathcal{Z}$ denote the set of all the integers (positive, negative, and 0), and define
$$S = \{nx + my : x \in \mathcal{Z}, \, y \in \mathcal{Z}\}.$$

Since $S$ clearly contains positive integers, it must (by the well-ordering property) contain a smallest positive integer. Let that integer be $d = ns + mt$. We will now show that $d = \gcd(n, m)$. To show that $d$ divides $n$, we use the division lemma to obtain the representation

$$n = qd + r, \quad 0 \le r < d,$$

which is equivalent to

$$0 \le r = n - qd = n - q(ns + mt) = n(1 - qs) - mqt < d.$$

Hence $r \in S$ and, since $d$ is the smallest positive member of $S$, this implies that $r$ cannot be positive. Therefore $r = 0$ and so $d$ divides $n$. By a similar argument we show that $d$ also divides $m$. Thus, $d$ is a common divisor of $n$ and $m$. In addition, if $c$ is also a common divisor then it must also be a divisor of $d = ns + mt$. Hence, $d = \gcd(n, m)$.    □

To find $\gcd(n, m)$ we use *Euclid's algorithm* (an algorithm is a procedure for solving a problem in a finite number of well-defined steps). If $n > m$, Euclid's algorithm uses the division lemma to show that if $r$ is the remainder when $n$ is divided by $m$ then $\gcd(n, m) = \gcd(m, r)$. It then continues in this fashion until the value of the greatest common divisor is obvious. We illustrate by an example.

**Example 1.5a**  Suppose we want to find $d = \gcd(84, 544)$. Dividing 544 by 84 gives a quotient 6 and remainder 40; that is,

$$544 = 6 \cdot 84 + 40.$$

Therefore, any common divisor of 84 and 544 is also a common divisor of 84 and 40, and vice versa. As a result,

$$d = \gcd(84, 544) = \gcd(84, 40).$$

However,
$$84 = 2 \cdot 40 + 4,$$

implying that $d = \gcd(40, 4)$. But

$$40 = 10 \cdot 4 + 0,$$

yielding that $d = \gcd(4, 0) = 4$. □

**Definition**  Any integer $n \geq 2$ that has no divisors except for 1 and itself is said to be *prime*.

The following important theorem is known both as the *prime factorization theorem* and as the *fundamental theorem of arithmetic*.

**Theorem 1.5.1** (Fundamental Theorem of Arithmetic) *Every integer $n > 1$ is either a prime or a product of primes that, except for the order of the prime factors, is unique.*

*Proof.* We prove this by mathematical induction. The theorem is true when $n = 2$ (since 2 is prime), so assume it to be true for all integers from 2 through $n - 1$, and consider $n$. If $n$ is prime, then the theorem is true for $n$; if $n$ is not prime, then for integers $a \geq 2$ and $b \geq 2$ we have

$$n = ab.$$

But since $a$ and $b$ are both less than $n$, it follows from the induction hypothesis that each is the product of primes, thus showing that $n$ is also the product of primes.

To show that the prime factorization of $n$ is unique, suppose that

$$n = p_1 \cdot p_2 \cdots p_r = q_1 \cdot q_2 \cdots q_s,$$

where all the $p_i$ and $q_j$ are primes. Now, there are two cases.

*Case 1:* $p_i = q_j$ for some $i$ and $j$. In this case, dividing the LHS of the preceding equation by $p_i$ and the RHS by $q_j$ yields

$$\prod_{k \neq i} p_k = \prod_{k \neq j} q_k.$$

But since $\prod_{k \neq i} p_k < n$, it follows from the induction hypothesis that its prime factorization must be unique. Hence $r = s$ and so $p_1, \ldots, p_r$ is just a rearrangement of $q_1, \ldots, q_s$. Thus, the prime factorization for $n$ is unique in Case 1.

*Case 2:* $p_i \neq q_j$ for all $i$ and $j$. Suppose that $p_1 > q_1$ (the argument is similar when the inequality is reversed). Let

$$k = (p_1 - q_1) p_2 \cdots p_r$$

and note that

$$
\begin{aligned}
k &= p_1 p_2 \cdots p_r - q_1 p_2 \cdots p_r \\
&= q_1 \cdot q_2 \cdots q_s - q_1 p_2 \cdots p_r \\
&= q_1 (q_2 \cdots q_s - p_2 \cdots p_r).
\end{aligned}
$$

Thus, $q_1$ divides $k$. Since $k/q_1$ and $p_1 - q_1$ are both integers less than $n$, it follows from the induction hypothesis that they can each be written as a product of primes – say,

$$\frac{k}{q_1} = t_1 \cdots t_b \quad \text{and} \quad p_1 - q_1 = s_1 \cdots s_m.$$

Therefore,

$$k = s_1 \cdots s_m \cdot p_2 \cdots p_r = q_1 \cdot t_1 \cdots t_b.$$

However, since $k < n$, it follows from the induction hypothesis that the preceding prime factorizations must simply be rearrangements of each other. As a result, $q_1$ must equal one of the $s_j$ or one of the $p_j$. But $q_1$ cannot equal any of the $s_j$, for they are the factors of $p_1 - q_1$ and $q_1$ cannot divide $p_1 - q_1$ (since $p_1$ and $q_1$ are unequal primes). Therefore, $q_1$ must equal one of the $p_j$, which shows that Case 2 cannot occur. The result has already been established for Case 1, so the proof is complete. □

The prime factorization theorem yields an elegant proof of the result that $\sqrt{n}$ is irrational when $n$ is not a perfect square.

**Definition**  The positive integer $n$ is said to be a *perfect square* if $n = k^2$ for some integer $k$.

**Corollary 1.5.1**  *For any positive integer $n$, $\sqrt{n}$ is irrational unless $n$ is a perfect square.*

**Proof.**  Suppose that, for integers $a$ and $b$,

$$\sqrt{n} = \frac{a}{b}.$$

Squaring both sides shows that

$$a^2 = nb^2.$$

It follows by the uniqueness of prime factors that, in the prime factorization of $a^2$, each distinct prime factor appears an even number of times; this holds likewise for the distinct prime factors of $b^2$. But this means that each of the distinct prime factors of $n$ also appears an even number of times in its prime factorization – for otherwise there would be a prime factor of $nb^2$ that appears an odd number of times in its prime factorization, which would contradict the uniqueness of the prime factorization of $a^2$. Therefore, $n$ is a perfect square. □

In fact, an almost identical proof can be use to prove a similar result for $r$th roots.

**Corollary 1.5.2**  *For positive integers $n$ and $r$, $n^{1/r}$ is irrational unless $n = k^r$ for some integer $k$.*

## 1.6    Exercises

**Exercise 1.1**   Show that the following statements are equivalent:

(a) $A \subset B$;
(b) $AB = A$;
(c) $A \cup B = B$.

**Exercise 1.2**   Suppose that $A$ and $B$ are subsets of the universal set $\mathcal{U}$. Find simpler expressions for the following sets:

(a) $A \cup A^c$;
(b) $AA^c$;
(c) $(A^c)^c$;
(d) $A \cup A^c B$;
(e) $AB \cup A^c B$.

**Exercise 1.3**   Prove *DeMorgan's laws*,

(a) $(A \cup B)^c = A^c B^c$ and
(b) $(AB)^c = A^c \cup B^c$.

**Exercise 1.4**   Verify that

$$AB \cup C = (A \cup C)(B \cup C).$$

**Exercise 1.5**   If $S = \bigcup_{j=1}^{k} S_j$, give a sufficient condition so that, for any numbers $x_i$ $(i \in S)$,

$$\sum_{i \in S} x_i = \sum_{j=1}^{k} \sum_{i \in S_j} x_i.$$

**Exercise 1.6**   Evaluate $\sum_{i=1}^{9} x_i + \sum_{i=1}^{10} y_i$ when $x_i = i^3 + 4$ and $y_i = 6 - i^3$.

**Exercise 1.7**   If $x_{i,j} = i + j^2$, find $\sum_{i=1}^{4} \sum_{j=1}^{3} x_{i,j}$.

**Exercise 1.8**  Simplify

$$\sum_{i=1}^{n}\sum_{j=1}^{i}\frac{i}{j}.$$

**Exercise 1.9**  Evaluate

$$\sum_{i=2}^{4}\sum_{j=i}^{2i}(i^2 - 2ij).$$

**Exercise 1.10**  Evaluate

$$\sum_{i=1}^{2}\sum_{j=1}^{4-i}\frac{j^2}{i}.$$

**Exercise 1.11**  Evaluate

$$\sum_{j=1}^{5}\sum_{i=j}^{5}\frac{1}{i}.$$

**Exercise 1.12**  Find an expression for $\sum_{i=1}^{n} ix^i$.
*Hint:* Start with the identity

$$\sum_{i=1}^{n} ix^i = \sum_{i=1}^{n}\sum_{j=1}^{i} x^i.$$

**Exercise 1.13**  With $a_i = i + 1$ and $x_i = (-1)^i$, evaluate:

(a) $\sum_{i=0}^{3} a_i x_i$;

(b) $\prod_{i=0}^{3} a_i x_i$;

(c) $\sum_{i=0}^{3}(a_i)^{x_i}$.

**Exercise 1.14**  Prove *Bernoulli's inequality*,

$$(1 + x)^n \geq 1 + nx, \quad x > -1.$$

**Exercise 1.15**  Show that, for any integer $n \geq 1$, $5^n - 1$ is divisible by 4.

**Exercise 1.16**  Show that, for any integer $n \geq 1$, $7^n - 1$ is divisible by 6.

**Exercise 1.17**  Show that, for any integer $n > 1$, $n^3 - n$ is divisible by 3.

**Exercise 1.18**  Give a direct argument to show that

$$\frac{1}{1 \cdot 2} + \frac{1}{2 \cdot 3} + \frac{1}{3 \cdot 4} + \cdots + \frac{1}{n(n+1)} = \frac{n+1}{n+2}.$$

*Hint:* What is $\frac{1}{i} - \frac{1}{i+1}$?

**Exercise 1.19**  Show that

$$\sum_{i=1}^{n} i^2 = \frac{n(n+1)(2n+1)}{6}, \quad n \geq 1.$$

**Exercise 1.20**  Show that

$$\sum_{i=1}^{n} i^3 = \left( \sum_{i=1}^{n} i \right)^2, \quad n \geq 1.$$

**Exercise 1.21**  Find a general formula for $1 + 3 + 5 + \cdots + (2n - 1)$ and then prove it to be true by induction.

**Exercise 1.22**  Find a general formula for $\sum_{i=1}^{n} i/(i+1)!$ and then prove it by induction. In the preceding expression, $j!$ (called $j$ *factorial*) is defined by

$$j! = \prod_{i=1}^{j} i = 1 \cdot 2 \cdot 3 \cdots j.$$

**Exercise 1.23**  Prove by mathematical induction that

$$1 + \sum_{i=1}^{n} ii! = (n+1)!, \quad n \geq 1.$$

Now, prove the preceding directly by writing

$$\sum_{i=1}^{n} ii! = \sum_{i=1}^{n}(i+1-1)i! = \sum_{i=1}^{n}(i+1)! - \sum_{i=1}^{n} i!.$$

**Exercise 1.24** Let $x_1 = 1$ and

$$x_{n+1} = 2x_n + 1, \quad n \geq 1.$$

Find and prove a general formula for $x_n$.

**Exercise 1.25** Let $x_1 = 3$ and

$$x_{n+1} = x_n(x_n + 2), \quad n \geq 1.$$

Find and prove a general formula for $x_n$.

**Exercise 1.26** Let $N(n)$ denote the number of distinct sequences of 0s and 1s in which no two adjacent values are both 1. Evaluate $N(2)$, $N(3)$, $N(4)$, $N(5)$. Propose an expression for $N(n)$ and then prove your conjecture by mathematical induction.

**Exercise 1.27** From a set of $m + 1$ distinct integers, show that we can choose two whose difference is a multiple of $m$.

**Exercise 1.28** Suppose that you have an unlimited supply of 10- and 13-cent stamps. Show that you can exactly total any postage amount greater than or equal to 108 cents.

**Exercise 1.29** Let $A$ be a subset of a universal set $\mathcal{U}$. The *characteristic function* of $A$, denoted by $f_A$, is defined by

$$f_A(x) = \begin{cases} 1 & \text{if } x \in A, \\ 0 & \text{if } x \notin A. \end{cases}$$

Prove the following:

(a) $f_{A^c}(x) = 1 - f_A(x)$;
(b) $f_{A \cup B}(x) = f_A(x) + f_B(x) - f_{AB}(x)$;
(c) $f_{AB}(x) = f_A(x) f_B(x)$;

(d) $f_{A \cup B}(x) - \max(f_A(x), f_B(x))$;

(e) $f_{AB}(x) = \min(f_A(x), f_B(x))$.

When is $f_A \leq f_B$?

**Exercise 1.30**   If $f$ and $g$ are both increasing functions on the integers, state whether (a) $f + g$ and (b) $fg$ are necessarily increasing. What if both $f$ and $g$ are nonnegative? (That is, what if $f(i) \geq 0$ and $g(i) \geq 0$ for all integers $i$?)

**Exercise 1.31**   Show that, if $f$ is convex, then $-f$ is concave (and vice versa).

**Exercise 1.32**   Prove Corollary 1.4.1.

**Exercise 1.33**   Show that the following function, having domain $\{0, 1, \ldots, n-1\}$, is convex:

$$f(j) = \frac{1}{n-j}.$$

**Exercise 1.34**   Show that $f(i) = e^i$ $(i = 0, 1, \ldots)$ is convex.

**Exercise 1.35**   Draw a graph of a convex function that:

(a)  is increasing;

(b)  first decreases and then increases;

(c)  first increases and then decreases.

**Exercise 1.36**   Repeat the preceding problem for a concave function.

**Exercise 1.37**   Show that if $f$ and $g$ are both nonnegative, increasing, and convex functions, then $fg$ is also convex.

**Exercise 1.38**   Let $f$ and $g$ be functions on the integers. With $h = fg$, show that

$$\Delta h(i) = f(i)\Delta g(i) + g(i-1)\Delta f(i).$$

**Exercise 1.39**   Find gcd(356, 10148).

**Exercise 1.40**    Find:

(a)  gcd(12321, 8658)
(b)  gcd(132, 473).

**Exercise 1.41**    If $\gcd(n, m) = 1$ then we say that $n$ and $m$ are *relatively prime*. Show that $n$ and $m$ are relatively prime if and only if there are integers $a$ and $b$ such that

$$an + bm = 1.$$

**Exercise 1.42**    If $n = 7^3 \cdot 5^4 \cdot 3^5$ and $m = 105 \cdot 10^5$, find $\gcd(n, m)$.

**Exercise 1.43**    Show that if $p$ is prime and is a divisor of $ab$ then it must also be a divisor of $a$ or $b$.

**Exercise 1.44**    Prove Corollary 1.5.2.

**Exercise 1.45**    Suppose that player $i$ had the most wins of any competitior in a round-robin tournament (see Example 1.3g). Show that, for every other player $j$, either $i$ beat $j$ or, for some $k$, $i$ beat $k$ and $k$ beat $j$.

# 2. Combinatorial Analysis

## 2.1    Introduction

Many problems can be solved simply by counting the number of different ways that a certain event can occur. In this chapter we show how one can efficiently do the counting in a variety of situations. In Section 2.2 we present the basic principle of counting, which is easily derived and extremely useful. Permutations are considered in Section 2.3 and combinations in 2.4. In Section 2.5, we consider the number of different solutions of certain integral linear equalities. A counting method based on inclusions and exclusions is presented in Section 2.6, and one based on deriving and solving recursion equations is presented in Section 2.7.

## 2.2    The Basic Principle of Counting

The following principle of counting will be basic to our work. Loosely put, it states that if an experiment consists of two parts, the first of which can result in any of $m$ possible outcomes and the second in any of $n$ possible outcomes, then there are a total of $mn$ possible outcomes of the experiment.

**Basic Principle of Counting**    Consider an experiment that consists of two phases. If the first phase can result in any of $m$ possible outcomes and if, for each outcome of the first phase, there are $n$ possible outcomes of the second phase, then there are a total of $mn$ possible outcomes of the experiment.

*Proof.* The basic principle can be proved by enumerating all possible outcomes of the experiment as follows:

$$(1, 1), \ (1, 2), \ \ldots, \ (1, n)$$

$$(2, 1), \ (2, 2), \ \ldots, \ (2, n)$$

$$\vdots$$

$$(m, 1), \ (m, 2), \ \ldots, \ (m, n),$$

where we say that the outcome is $(i, j)$ if the first phase of the experiment results in its $i$th possible outcome and the next phase then results in the $j$th of its possible outcomes. Thus, the set of possible outcomes can be represented in $m$ rows, each row containing $n$ outcomes, which proves the result.                                          $\square$

**Example 2.2a**   A women's group consists of twelve women, each of whom has three children. If one woman and one of her children are to be chosen, how many different choices are possible?

*Solution.*   By regarding the choice of the woman as the outcome of the first phase of the experiment and the subsequent choice of her child as the outcome of the second phase, we see from the basic principle that there are $12 \cdot 3 = 36$ possible choices.                           $\square$

When there are more than two phases to the experiment, the basic principle can be generalized as follows.

**Generalized Basic Principle of Counting**   Consider an experiment that consists of $r$ phases. If the first phase can result in any of $n_1$ possible outcomes, and if for each outcome of the first phase there are $n_2$ possible outcomes of the second phase, and if for each of the possible outcomes of the first two phases there are $n_3$ possible outcomes of the third phase, and if $\ldots$, then there are a total of $n_1 \cdot n_2 \cdots n_r$ possible outcomes of the experiment.

**Example 2.2b**   A school planning committee consists of 3 freshmen, 4 sophomores, 4 juniors, and 5 seniors. If a subcommittee is to be chosen that consists of one person from each class, how many choices are possible?

*Solution.* The choice of the subcommittee can be considered as the outcome of a four-phase experiment, where phase 1 is the choice of the freshmen member of the subcommittee, phase 2 is the choice of the sophomore member, and so on. Hence, from the generalized version of the basic principle of counting, it follows that there a total of $3 \cdot 4 \cdot 4 \cdot 5 = 240$ possible subcommittees.                                                   □

**Example 2.2c**  How many different seven-place license numbers are possible if the first three places are to be occupied by letters and the final four by numbers? What if no repetition of letters or numbers is allowed?

*Solution.*  By the generalized basic principle, there are

$$26 \cdot 26 \cdot 26 \cdot 10 \cdot 10 \cdot 10 \cdot 10 = 175{,}760{,}000$$

possible license plates when repetitions are allowed. If repetitions are not allowed, then there are only

$$26 \cdot 25 \cdot 24 \cdot 10 \cdot 9 \cdot 8 \cdot 7 = 78{,}624{,}000.$$                          □

**Example 2.2d**  How many different functions on $n$ points are possible if each functional value is either 0 or 1?

*Solution.*  Let the points be denoted $1, 2, \ldots, n$. Since $f(i)$ must have one of two possible values, it follows from the generalized basic principle that there are a total of $2^n$ possible functions.                            □

## 2.3    Permutations

How many different *ordered* arrangements of the letters $a, b, c$ are possible? By direct enumeration we see that there are six – namely, *abc*, *acb*, *bac*, *bca*, *cab*, and *cba*. Each arrangement is known as a *permutation*. Thus we see that there are a total of six permutations of a set of three objects. This result could also have been obtained from the generalized basic counting principle, since the first object in the permutation is any of the three, the second can then be chosen from either of the remaining two, and the third object in the permutation is then the one that remains. Thus there are $3 \cdot 2 \cdot 1 = 6$ possible permutations.

Suppose now that we have $n$ objects. Reasoning in the same manner as before, we see that there are a total of

$$n(n-1)(n-2)\cdots 3\cdot 2\cdot 1 = n!$$

different permutations of $n$ objects.

**Example 2.3a**  An advanced mathematics class consists of six women and four men. An examination is given, and the students are ranked according to their performance. Assume that no two students obtain the same score.

(a) How many different rankings are possible?
(b) If the women are ranked just among themselves and the men among themselves, how many different rankings are possible?

*Solution.*  As each ranking corresponds to a particular ordered arrangement of the ten people, it follows that there are $10! = 3,628,800$ possible rankings. On the other hand, since there are $6!$ possible rankings of the women among themselves and $4!$ possible rankings of the men among themselves, it follows from the basic principle of counting that there are $6!\,4! = 17,280$ possible rankings in case (b). □

**Example 2.3b**  Smith has twelve books to put on a bookshelf. Of these, five are mathematics books, three are economics books, two are history books, and two are language books. Smith wants to arrange these books so that all the books dealing with the same subject are together on the shelf. How many different arrangements are possible?

*Solution.*  There are $5!\,3!\,2!\,2!$ arrangements such that the mathematics books are first in line, the economics books are second, the history books are third, and the language books are last. Similarly, for each possible ordering of the subjects there are this number of different possible arrangements. Hence, as there are $4!$ possible orderings of the subjects, we see that there are a total of $4!\,5!\,3!\,2!\,2! = 69,190$ possible arrangements. □

The next example concerns the number of different permutations of a set of objects when some of those objects are indistinguishable from one another.

**Example 2.3c**   How many different letter arrangements can be formed from the letters $R, A, B, B, A$?

**Solution.**  First note that there are 5! permutations of the letters $R, A_1, B_1, B_2, A_2$ when the two $B$s and the two $A$s are distinguished from each other. However, consider any one of these permutations – say, $B_1, A_1, B_2, A_2, R$. If we now permute the $B$s among themselves and the $A$s among themselves, then the resulting arrangement will still be of the form $BABAR$. That is, each of the 2! 2! permutations

$$B_1 \ A_1 \ B_2 \ A_2 \ R$$
$$B_2 \ A_1 \ B_1 \ A_2 \ R$$
$$B_1 \ A_2 \ B_2 \ A_1 \ R$$
$$B_2 \ A_2 \ B_1 \ A_1 \ R$$

is still of the form $BABAR$. Hence, there are $5!/2! \, 2! = 30$ possible letter arrangements.  □

In general, the same reasoning as used in Example 2.3c shows that there are

$$\frac{n!}{n_1! \, n_2! \cdots n_r!}$$

different permutations of $n$ items, of which $n_1$ are alike, $n_2$ are alike, ..., and $n_r$ are alike, where $n = \sum_{i=1}^{r} n_i$.

**Example 2.3d**   Suppose that we have a large collection of beads where each bead is one of $n$ different colors, and suppose that we want to use the beads to make a necklace consisting of $p$ beads. How many different necklaces are there that use beads of at least two different colors when $p$ is a prime number?

**Solution.**  A necklace can be made by first lining up $p$ beads in a row and then running a string through holes drilled in the center of the beads; the ends of the string are then connected to form the necklace (see Figure 2.1).

There are $n^p$ different possible color arrangements of $p$ beads in a row. Since $n$ of these arrangements will consist of beads of the same

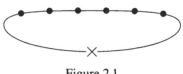

Figure 2.1

color, it follows that there are $n^p - n$ different multicolor arrangements. However, some of these arrangements result in the same necklace. To see how many different necklaces there are, consider any linear arrangement of beads – say,

$$r, b, b, g, r, \ldots, r,$$

where $r$ stands for a red, $b$ for a blue, and $g$ for a green bead. If we make the transformation of moving the first bead to the end of the line to obtain the new arrangement

$$b, b, g, r, \ldots, r, r,$$

then these two arrangements will result in the same necklace when the ends of the string are connected. Now consider how many of these transformations (obtained by moving the front end bead to the back of the line) we can make before the original color arrangement is restored. Let $m$ be the smallest number of such transformations, and note that $m > 1$ (since the string of beads is multicolored). It is easy to see that the original color arrangement will be restored after (and only after) $km$ transformations, for any positive integer $k$. However, because there are $p$ beads in a string, it follows that the original arrangement is also obtained after $p$ transformations. Thus, $p$ must be an integral multiple of $m$, which implies (since $p$ is prime) that $m = p$. Hence we can partition the $n^p - n$ linear color arrangements into groups of size $p$, where each arrangement in a group gives rise to the same necklace. In addition, we claim that necklaces from different groups are distinct. To see this, note that two necklaces are identical when a rotation of one of the necklaces transforms it into the other. However, since rotating a necklace is equivalent to successive transformations of moving the first bead to the end of the line before connecting its strings, we can conclude that necklaces in different groups are distinct. This shows that there are a total of $(n^p - n)/p$ different possible necklaces.     □

Since the number of possible necklaces of Example 2.3g must be an integer, we obtain a famous result in number theory as follows.

**Fermat's Little Theorem**     *If p is a prime number, then $n^p - n$ is divisible by p for any integer $n > 1$.*

## 2.4     Combinations

We are often interested in determining the number of different groups of $r$ items that can be formed from a total of $n$ items. For instance, how many different groups of size three can be formed from the set of five letters A, B, C, D, E? To determine this number, note that there are five different choices of the first member of the group, then four different choices for the second, and then three different choices for the third. Hence, there are $5 \cdot 4 \cdot 3$ different choices when the order in which the items are selected is considered relevant. However, each group of three – say, the group consisting of items A, B, and C – will be counted 3! times (since each of the orderings ABC, ACB, BAC, BCA, CAB, CBA will be counted), so it follows that the total number of groups that can be formed is

$$\frac{5 \cdot 4 \cdot 3}{3 \cdot 2 \cdot 1} = 10.$$

In general, as $n(n - 1) \cdots (n - r + 1)$ represents the number of different ways that a group of $r$ items can be selected from $n$ items when the order of selection is considered relevant, and as each group of $r$ items will be counted $r!$ times in this count, it follows that the number of different groups of $r$ items that can be formed from a set of $n$ items is

$$\frac{n(n - 1) \cdots (n - r + 1)}{r!} = \frac{n!}{(n - r)! \, r!}.$$

**Definition**     For $0 \leq r \leq n$, define $\binom{n}{r}$ by

$$\binom{n}{r} = \frac{n!}{(n - r)! \, r!}.$$

(By convention, 0! is defined to equal 1 and so $\binom{n}{0} = \binom{n}{n} = 1$. Also, we take $\binom{n}{r}$ to equal 0 when $r > n$.)

The notation $\binom{n}{r}$ thus represents the number of different groups of $r$ items that can be chosen from a set of $n$ items when the order in which the items are selected is not considered relevant.

**Example 2.4a** A committee of four is to be chosen from a group of 16 people. How many different committees are possible?

*Solution.* There are

$$\binom{16}{4} = \frac{16 \cdot 15 \cdot 14 \cdot 13}{4 \cdot 3 \cdot 2 \cdot 1} = 1{,}820$$

different committees. ☐

**Example 2.4b** From a group of six women and five men, how many different committees consisting of three women and two men can be formed? How many can be formed if two of the women do not want to serve together?

*Solution.* Since there are $\binom{6}{3} = 20$ possible groups of three women and $\binom{5}{2} = 10$ possible groups of two men, it follows from the basic principle of counting that there are 200 possible committees of three women and two men.

Supppose now that two of the women do not want to serve together. There are $\binom{4}{3} = 4$ groups of three women not containing either of them and $\binom{2}{1}\binom{4}{2} = 12$ groups of three women that contain exactly one of them, so it follows that there are a total of 16 groups of three women that do not contain both of these women. Since there are 10 different groups of two men, it follows from the basic principle that there are 160 possible committees in this case. ☐

**Example 2.4c** A group of $n$ components consists of $m$ that are defective (D) and $n - m$ that are functional (F). These components are to be lined up in such a fashion that no two defectives are next to each

---

$$- \; + \; - \; + \; - \; + \; - \; + \; - \; + \; -$$

$+ =$ functional
$- =$ place for at most one defective

Figure 2.2

---

other. How many linear arrangements are possible if we do not distinguish among the functional components nor among the defective ones? For instance, when $n = 4$ and $m = 2$ there are three such arrangements: FDFD, DFDF, and DFFD.

*Solution.* Imagine that the $n - m$ functional components are lined up among themselves. If no two defectives are to be consecutive, then the spaces between the functional components cannot contain more than a single defective component. Therefore, we must choose $m$ of the $n - m + 1$ possible positions between the $n - m$ functional components (see Figure 2.2) and place a defective component in each position chosen. As a result, there are $\binom{n-m+1}{m}$ possible orderings.    □

**Example 2.4d**    The vector $(x_1, x_2, \ldots, x_n)$ is said to be a *word* of length $n$ from the alphabet $R = \{1, \ldots, r\}$ if each $x_i \in R$. The distance between the words $\mathbf{x} = (x_1, \ldots, x_n)$ and $\mathbf{y} = (y_1, \ldots, y_n)$ is defined to equal the number of indices $i$ for which $x_i \neq y_i$. That is, with $\rho(\mathbf{x}, \mathbf{y})$ equal to the distance between $\mathbf{x}$ and $\mathbf{y}$, we have

$$\rho(\mathbf{x}, \mathbf{y}) = \sum_{i=1}^{n} I(x_i, y_i),$$

where $I(x, y)$ is 0 when $x = y$ and is 1 otherwise.

(a) How many $n$-letter words are there?
(b) How many $n$-letter words are within a distance $k$ of a specified word $\mathbf{x}$? That is, how many $n$-letter words $\mathbf{y}$ are such that $\rho(\mathbf{x}, \mathbf{y}) \leq k$?

*Solution.* Since each letter can be any of $r$ possibilities, it follows from the generalized basic principle of counting that there are $r^n$ distinct $n$-letter words. To answer part (b), consider the number of $n$-letter words that are a distance $i$ from **x**. There are $\binom{n}{i}$ choices of the positions of the $i$ letters of **x** that are to be changed and then $r - 1$ choices for each changed letter, so it follows that there are $\binom{n}{i}(r-1)^i$ such words. Hence, the solution to part (b) is

$$\sum_{i=1}^{k}\binom{n}{i}(r-1)^i. \qquad \square$$

The following is a useful combinatorial identity:

$$\binom{n+1}{i} = \binom{n}{i-1} + \binom{n}{i}. \qquad (2.1)$$

Equation (2.1) can be proved analytically or by the following combinatorial argument. Consider a group of $n+1$ objects and fix attention on one of them, call it item number 1. Note that there are $\binom{n}{i-1}$ subgroups of size $i$ that contain item number 1 (since such a subgroup is obtained by choosing $i - 1$ of the remaining $n$ items). Similarly, there are $\binom{n}{i}$ subgroups of size $i$ that do not contain item 1 (since such a subgroup is obtained by choosing $i$ of the remaining $n$ items). As there are a total of $\binom{n+1}{i}$ subgroups of size $i$, equation (2.1) follows.

The quantity $\binom{n}{r}$ is often called a *binomial coefficient* because of its prominence in the binomial theorem.

**Binomial Theorem**

$$(x + y)^n = \sum_{i=0}^{n}\binom{n}{i}x^i y^{n-i}. \qquad (2.2)$$

We will present two proofs of the binomial theorem. The first uses mathematical induction, whereas the second is based on combinatorial considerations.

*Induction Proof.* When $n = 1$, equation (2.2) reduces to

$$x + y = \binom{1}{0}x^0 y^1 + \binom{1}{1}x^1 y^0 = y + x.$$

So assume that the equation is valid for $n$. Then,

$$(x + y)^{n+1} = (x + y)(x + y)^n$$

$$= (x + y) \sum_{k=0}^{n} \binom{n}{k} x^k y^{n-k}$$

$$= \sum_{k=0}^{n} \binom{n}{k} x^{k+1} y^{n-k} + \sum_{k=0}^{n} \binom{n}{k} x^k y^{n-k+1}.$$

Letting $i = k + 1$ in the first sum and $i = k$ in the second yields

$$(x + y)^{n+1} = \sum_{i=1}^{n+1} \binom{n}{i-1} x^i y^{n-i+1} + \sum_{i=0}^{n} \binom{n}{i} x^i y^{n-i+1}$$

$$= x^{n+1} + \sum_{i=1}^{n} \left[ \binom{n}{i-1} + \binom{n}{i} \right] x^i y^{n-i+1} + y^{n+1}$$

$$= x^{n+1} + \sum_{i=1}^{n} \binom{n+1}{i} x^i y^{n-i+1} + y^{n+1}$$

$$= \sum_{i=0}^{n+1} \binom{n+1}{i} x^i y^{n+1-i},$$

where the next-to-last equality follows from (2.1). By mathematical induction, the binomial theorem is proved.    □

*Combinatorial Proof.* Consider the product

$$(x_1 + y_1)(x_2 + y_2) \cdots (x_n + y_n);$$

its expansion consists of the sum of $2^n$ terms, where each term is the product of $n$ factors. Further, each of the $2^n$ terms in the sum will have as a factor exactly one of $x_j$ or $y_j$ for each $j = 1, \ldots, n$. For instance,

$$(x_1 + y_1)(x_2 + y_2) = x_1 x_2 + x_1 y_2 + y_1 x_2 + y_1 y_2.$$

Now, how many of the $2^n$ terms in the sum will have as factors $k$ of the $x_j$ and $n - k$ of the $y_j$? Because each term consisting of $k$ of the $x_j$ and $n - k$ of the $y_j$ corresponds to a choice of a group of size $k$ from the terms $x_1, x_2, \ldots, x_n$, it follows that there are $\binom{n}{k}$ such terms. Thus, letting $x_j = x$ and $y_j = y$ ($j = 1, \ldots, n$), we see that

$$(x + y)^n = \sum_{k=0}^{n} \binom{n}{k} x^k y^{n-k}$$

and the result is proved. □

**Example 2.4e**  Expand $(x + v)^3$

**Solution.**

$$(x + y)^3 = \binom{3}{0} x^0 y^3 + \binom{3}{1} x^1 y^2 + \binom{3}{2} x^2 y^1 + \binom{3}{3} x^3 y^0$$
$$= y^3 + 3xy^2 + 3x^2 y + x^3.$$  □

**Example 2.4f**  How many subsets are there of a set consisting of $n$ elements?

**Solution.**  Since there are $\binom{n}{i}$ subsets of size $i$, the answer is

$$\sum_{i=0}^{n} \binom{n}{i} = \sum_{i=0}^{n} \binom{n}{i} 1^i 1^{n-i} = (1 + 1)^n = 2^n.$$

The result could also have been obtained by assigning to each element in the set either the value 0 or the value 1. To each assignment of values there corresponds, in an one-to-one fashion, a subset; namely, that subset consisting of all elements having value 1. Since there are $2^n$ possible assignments, the result follows. (It should be noted that we have included as a subset the null set, which consists of no elements. Hence, the number of *nonempty* subsets is $2^n - 1$.)  □

## 2.5   Counting the Number of Solutions

There are $r^n$ possible outcomes when $n$ distinguishable balls are to be distributed into $r$ distinguishable urns. This follows because each ball may be put into any of $r$ possible urns. However, suppose now that the

$$\underset{\wedge}{\bigcirc}\underset{\wedge}{\bigcirc}\underset{\wedge}{\bigcirc}\underset{\wedge}{\bigcirc}\underset{\wedge}{\bigcirc}\underset{\wedge}{\bigcirc}\underset{\wedge}{\bigcirc}\underset{\wedge}{\bigcirc}\underset{\wedge}{\bigcirc}\underset{\wedge}{\bigcirc}\underset{\wedge}{\bigcirc}\bigcirc$$

choose $r - 1$ of the spaces $\wedge$

Figure 2.3

balls are indistinguishable from each other and consider how many outcomes there are in this case. Since the balls are indistinguishable, it follows that the outcome of the experiment of putting the $n$ balls into $r$ distinguishable urns is the vector $(x_1, x_2, \ldots, x_r)$, where $x_i$ is the number of balls that are put into urn $i$. Thus, the problem reduces to finding the number of nonnegative integer-valued vectors $(x_1, \ldots, x_r)$ such that

$$x_1 + x_2 + \cdots + x_r = n.$$

In order to determine this number, start by considering the number of *positive* integer-valued solutions of the preceding equation. Toward this end, imagine that we have $n$ indistinguishable items lined up and that we want to divide them into $r$ nonempty groups. To accomplish this, we can select $r - 1$ of the $n - 1$ spaces between adjacent items as our dividing points (see Figure 2.3).

For instance, let $n = 6$ and $r = 3$, and choose the two dividing points as follows.

$$\bigcirc\bigcirc\underset{\wedge}{}\bigcirc\bigcirc\bigcirc\underset{\wedge}{}\bigcirc$$

Then the vector obtained is $x_1 = 2$, $x_2 = 3$, $x_3 = 1$. As there are $\binom{n-1}{r-1}$ possible selections of the dividing points, we obtain the following proposition.

**Proposition 2.5.1**   *There are $\binom{n-1}{r-1}$ distinct positive integer-valued vectors $(x_1, \ldots, x_n)$ satisfying*

$$x_1 + x_2 + \cdots + x_r = n.$$

To obtain the number of nonnegative (as opposed to positive) solutions, note that the number of nonnegative solutions of $x_1 + x_2 + \cdots + x_r = n$ is the same as the number of positive solutions of $y_1 + y_2 + \cdots + y_r =$

$n+r$ (this may be seen by letting $y_i = x_i + 1$ for $i = 1, \ldots, r$). Therefore, from Proposition 2.5.1 we obtain the following corollary.

**Corollary 2.5.1**    *There are $\binom{n+r-1}{r-1}$ distinct nonnegative integer-valued vectors $(x_1, \ldots, x_n)$ satisfying*

$$x_1 + x_2 + \cdots + x_r = n.$$

**Example 2.5a**    How many distinct nonnegative integer-valued solutions of $x_1 + x_2 + x_3 = 2$ are possible?

*Solution.*    There are $\binom{4}{2} = 6$ solutions: $(0, 0, 2)$, $(0, 2, 0)$, $(2, 0, 0)$, $(1, 1, 0)$, $(1, 0, 1)$, and $(0, 1, 1)$. $\qquad\qquad\qquad\qquad\qquad\qquad\qquad\qquad$ □

**Example 2.5b**    An investor has \$25,000 to invest among four possible investments. Each investment must be in units of a thousand dollars. If the entire amount is to be invested, how many different investment strategies are possible? What if all the money need not be invested?

*Solution.*    If we let $x_i$ denote the number of thousands of dollars to be put into investment $i$ ($i = 1, 2, 3, 4$) then, when all the money needs to be invested, these quantities must satisfy

$$x_1 + x_2 + x_3 + x_4 = 25.$$

Hence, by Corollary 2.5.1, there are $\binom{28}{3} = 3{,}276$ possible investment strategies. If not all the money needs to be invested, then (letting $x_5$ denote the amount not invested) it follows that an investment strategy is a nonnegative integer-valued vector $(x_1, \ldots, x_5)$ satisfying

$$x_1 + x_2 + x_3 + x_4 + x_5 = 25.$$

By Corollary 2.5.1, there are now $\binom{29}{4} = 23{,}751$ possible investment strategies. $\qquad\qquad\qquad\qquad\qquad\qquad\qquad\qquad\qquad\qquad$ □

## 2.6    The Inclusion–Exclusion Identity

For a finite set $A$, let $N(A)$, called the *cardinality* of the set $A$ (and often denoted as $|A|$), denote the number of elements in $A$. The following identity relates the number of elements that are in either of two sets to the number in each and the number in both:

$$N(A \cup B) = N(A) + N(B) - N(AB). \tag{2.3}$$

The preceding is easily established by noting that an element that is in exactly one of the sets $A$ and $B$ is counted once on both sides of the identity, whereas an element that is in both $A$ and $B$ adds 1 to the count of the LHS and $1 + 1 - 1 = 1$ to the count of the RHS.

An expression for the number of elements in the union of three sets can be obtained by using (2.3) as follows:

$$N(A \cup B \cup C) = N((A \cup B) \cup C)$$
$$= N(A \cup B) + N(C) - N((A \cup B)C)$$
$$= N(A) + N(B) - N(AB) + N(C) - N(AC \cup BC)$$
$$= N(A) + N(B) - N(AB) + N(C)$$
$$\quad - N(AC) - N(BC) + N(ACBC)$$
$$= N(A) + N(B) + N(C) - N(AB)$$
$$\quad - N(AC) - N(BC) + N(ABC).$$

By continuing in this fashion, it is not difficult to discern the general pattern, which is given in the following proposition.

**Proposition 2.6.1** (Inclusion–Exclusion Identity)

$$N\left(\bigcup_{j=1}^{n} A_j\right)$$
$$= \sum N(A_j) - \sum\sum N(A_j A_k)$$
$$+ \sum\sum\sum N(A_j A_k A_s) - \cdots + (-1)^{n+1} N(A_1 A_2 \cdots A_n),$$

*where the signs of the successive sums continually change. The first sum is over all the n values of $j$; the second is over all the $\binom{n}{2}$ pairs $j < k$; the third is over all the $\binom{n}{3}$ pairs $j < k < s$; and so on.*

*Proof.* The inclusion–exclusion identity can be proved either by mathematical induction or by the following argument. Consider any element that is in exactly $r$ of the sets $A_j$, $j = 1, \ldots, n$. Note that it is also in exactly $\binom{r}{2}$ of the sets $A_j A_k$ ($j < k$) and in exactly $\binom{r}{3}$ of the sets $A_j A_k A_s$ ($j < k < s$) and so on. Hence, whereas such an element will be counted once in the LHS of the identity, it will add

$$r - \binom{r}{2} + \binom{r}{3} - \cdots = \sum_{j=1}^{r} \binom{r}{j}(-1)^{j+1}$$

to the count on the RHS. The result will then follow if we show that, for $r = 1, \ldots, n$,

$$1 = \sum_{j=1}^{r} \binom{r}{j}(-1)^{j+1}$$

or (equivalently) that

$$1 - \sum_{j=1}^{r} \binom{r}{j}(-1)^{j+1} = 0$$

or

$$1 + \sum_{j=1}^{r} \binom{r}{j}(-1)^{j+2} = 0,$$

which, since $(-1)^{j+2} = (-1)^{j}$, is equivalent to

$$\sum_{j=0}^{r} \binom{r}{j}(-1)^{j} = 0.$$

But the preceding follows from the binomial theorem, since

$$0 = (-1+1)^r = \sum_{j=0}^{r} \binom{r}{j}(-1)^{j}. \qquad \square$$

**Example 2.6a**   A total of 34 members of a club play tennis, 26 play squash, and 15 play badminton. Furthermore, 20 play both tennis and squash, 11 play both tennis and badminton, 8 play both squash and badminton, and 4 play all three sports. How many club members play at least one of these sports?

***Solution.***   Let $T$, $S$, $B$ denote the numbers that play tennis, squash, and badminton, respectively. Then

$$N(T \cup S \cup B) - N(T) + N(S) + N(B) - N(TS)$$
$$- N(TB) - N(SB) + N(TSB)$$
$$= 34 + 26 + 15 - 20 - 11 - 8 + 4$$
$$= 40.$$

Thus, a total of 40 club members play tennis or squash or badminton.

$\square$

**Example 2.6b**   Any permutation $i_1, i_2, \ldots, i_n$ of the numbers $1, 2, \ldots, n$ is said to be a *derangement* if $i_j \neq j$ for each $j = 1, \ldots, n$. That is, a derangement is a permutation in which none of the elements is in its normal position. How many derangements are there?

*Solution.*   Let us answer the preceding question by first determining the number of permutations that are *not* derangements. To do so, let $A_j$ ($j = 1, \ldots, n$) determine the set of all permutations $i_1, \ldots, i_j, \ldots, i_n$ of $1, \ldots, n$ for which $i_j = j$. Since a permutation will not be a derangement if and only if it is in $A_j$ for some $j$, it follows that $N\left(\bigcup_{j=1}^{n} A_j\right)$ is the number of permutations that are not derangements. To determine its value, we will use the inclusion–exclusion identity. Note first that there are $(n - i)!$ permutations that have $i$ of their positions specified; as a result, we see that

$$N(A_j) = (n - 1)!,$$
$$N(A_j A_k) = (n - 2)! \quad (j < k),$$
$$N(A_j A_k A_s) = (n - 3)! \quad (j < k < s),$$

and so on. Hence, from the inclusion–exclusion identity we obtain that the number of permutations that are not derangements is given by

$$N\left(\bigcup_{j=1}^{n} A_j\right) = \binom{n}{1}(n - 1)! - \binom{n}{2}(n - 2)! + \cdots + (-1)^{n+1}\binom{n}{n}0!$$

$$= \sum_{i=1}^{n}\binom{n}{i}(n - i)!\,(-1)^{i+1}$$

$$= n!\sum_{i=1}^{n}\frac{(-1)^{i+1}}{i!}.$$

The total number of permutations is $n!$, so it follows that there are

$$n! - n! \sum_{i=1}^{n} \frac{(-1)^{i+1}}{i!} = n!\left(1 - \sum_{i=1}^{n} \frac{(-1)^{i+1}}{i!}\right)$$

$$= n!\left(1 + \sum_{i=1}^{n} \frac{(-1)^{i+2}}{i!}\right)$$

$$= n! \sum_{i=0}^{n} \frac{(-1)^{i}}{i!}$$

derangements. □

**Example 2.6c** *Coupon Collecting Problem*     Suppose that one is to collect $m$ coupons, each of which is one of $n$ possible types. Let $x_i$ denote the type of the $i$th coupon collected, and say that $(x_1, \ldots, x_m)$ is the outcome of the experiment. How many outcomes are there for which at least one coupon of each type is collected?

*Solution.*   Let $A_j$ denote the set of outcomes for which none of the $x_i$ are equal to $j$. That is, $A_j$ is the set of outcomes in which a type-$j$ coupon is not obtained. Then $N\left(\bigcup_{j=1}^{n} A_j\right)$ denotes the number of outcomes in which a complete set is not obtained. We have

$$N(A_j) = (n-1)^m,$$

$$N(A_j A_k) = (n-2)^m \quad (j < k),$$

$$N(A_j A_k A_s) = (n-3)^m \quad (j < k < s),$$

and so on, which follows because there are $(n-i)^m$ outcomes in which the $m$ coupons are restricted to come from only $n-i$ types, $i \geq 1$. Hence, from the inclusion–exclusion identity we obtain that

$$N\left(\bigcup_{j=1}^{n} A_j\right) = \sum_{i=1}^{n} \binom{n}{i}(n-i)^m(-1)^{i+1}$$

$$= \sum_{j=0}^{n-1} \binom{n}{n-j} j^m (-1)^{n-j+1} \quad \text{(by letting } j = n - i)$$

$$= \sum_{j=0}^{n-1} \binom{n}{j} j^m (-1)^{n-j+1}.$$

There are a total of $n^m$ outcomes, so the number of outcomes in which a complete set is obtained is

$$n^m - N\left(\bigcup_{j=1}^{n} A_j\right) = n^m + \sum_{j=0}^{n-1} \binom{n}{j} j^m (-1)^{n-j+2}$$

$$= \sum_{j=0}^{n} \binom{n}{j} j^m (-1)^{n-j}. \qquad \square$$

When $m < n$, it is impossible to obtain a complete set in the coupon collecting problem, which gives rise to the following useful combinatorial identity.

**Corollary 2.6.1**    *If $m < n$, then*

$$\sum_{j=0}^{n} \binom{n}{j} j^m (-1)^{n-j} = 0.$$

## 2.7    Using Recursion Equations

Suppose that we want to determine the number of different ways a certain type of procedure can be performed with $n$ items. An approach that is often fruitful is to let $N(n)$ denote this quantity and then try to derive an expression for $N(n)$ in terms of the quantities $N(i)$, $i < n$. Starting with the value of $N(1)$, we can use this *recursion* equation to find $N(2)$, then $N(3)$, and so on. In so doing we are often able to discern a pattern that enables us to explicity find $N(n)$, and even when this is not possible the recursion approach is often a computationally efficient way of evaluating $N(n)$. We will illustrate by a series of examples. In Example 2.7a we solve the problem of Example 2.2d by the recursion approach.

**Example 2.7a**    How many subsets are there of $S = \{1, \ldots, n\}$?

**Solution.**    Let $N(n)$ denote the number of subsets of the set $\{1, \ldots, n\}$, and consider the relation between $N(n)$ and $N(n-1)$. To begin, note that the number of subsets of $\{1, \ldots, n\}$ is equal to the number of subsets that do not contain the element $n$ *plus* the number that do. The number of subsets that do not contain $n$ is clearly equal to $N(n-1)$, the number of subsets of the remaining $n-1$ elements. In addition, since any subset containing the outcome $n$ can be obtained by adding $n$ to a subset of the other $n-1$ elements, it also follows that there are $N(n-1)$ of these subsets. Hence we see that

$$N(n) = N(n-1) + N(n-1) = 2N(n-1), \quad n \geq 2.$$

Since $N(1) = 2$ (the two subsets are $\emptyset$ and $\{1\}$), we obtain that

$$N(1) = 2,$$
$$N(2) = 2N(1) = 2^2,$$
$$N(3) = 2N(2) = 2^3.$$

It is now easy to see (and prove formally by mathematical induction) that

$$N(n) = 2^n. \qquad \square$$

**Example 2.7b**   A "codeword" from the alphabet $\{0, 1, 2, 3\}$ is said to be *legitimate* if it contains an even number of zeros. Thus, for instance, the codeword 31010 is legitimate whereas 01010 is not. How many $n$-letter codewords are legitimate?

*Solution.*   Let $N(n)$ denote the number of $n$-letter legitimate codewords. Let us determine how many of these $N(n)$ codewords start with each value in the alphabet. If the first value is 0, then the remaining $n-1$ values must constitute an $(n-1)$-letter codeword with an odd number of zeros. There are a total of $4^{n-1}$ different $(n-1)$-letter codewords, of which $N(n-1)$ contain an even number of zeros, so it follows that there are $4^{n-1} - N(n-1)$ that contain an odd number of zeros. Thus, there are $4^{n-1} - N(n-1)$ $n$-letter legitimate codewords that start with 0. Since the number of $n$-letter legitimate codewords that start with 1 (or with 2 or 3) is equal to the number of $(n-1)$-letter legitimate codewords, we see that

$$N(n) = 4^{n-1} - N(n-1) + 3N(n-1)$$
$$= 4^{n-1} + 2N(n-1).$$

Using this expression recursively – first with $n - 2$, then $n = 3$, and so on – yields

$$N(2) = 4 + 2N(1),$$

$$N(3) = 4^2 + 2N(2) = 4^2 + 2 \cdot 4 + 2^2 N(1),$$

$$N(4) = 4^3 + 2N(3) = 4^3 + 2 \cdot 4^2 + 2^2 \cdot 4 + 2^3 N(1),$$

$$N(5) = 4^4 + 2N(4) = 4^4 + 2 \cdot 4^3 + 2^2 \cdot 4^2 + 2^3 \cdot 4 + 2^4 N(1).$$

It is easy to see (the formal proof is left as an exercise) the pattern: namely, that

$$\begin{aligned}
N(n) &= 4^{n-1} + 2 \cdot 4^{n-2} + 2^2 \cdot 4^{n-3} + \cdots + 2^{n-2} \cdot 4 + 2^{n-1} N(1) \\
&= 2^{2n-2} + 2^{2n-3} + \cdots + 2^n + 2^{n-1} N(1) \\
&= 2^n [1 + 2 + \cdots + 2^{n-2}] + 2^{n-1} N(1) \\
&= 2^n (2^{n-1} - 1) + 2^{n-1} N(1) \\
&= 2^{2n-1} - 2^n + 2^{n-1} N(1).
\end{aligned}$$

Since $N(1) = 3$, we see that

$$N(n) = 2^{2n-1} + 2^{n-1}.$$

An interesting combinatorial identity is obtained by noting that there are exactly $\binom{n}{r} 3^{n-r}$ codewords that contain $r$ zeros. This follows because there are $\binom{n}{r}$ locations for these zeros and each of the other $n - r$ digits can then be any of the other three values from the alphabet. Hence, from the preceding formula for $N(n)$, we obtain that

$$\sum_{i=0}^{[n/2]} \binom{n}{2i} 3^{n-2i} = 2^{2n-1} + 2^{n-1},$$

where $[n/2]$ denotes the largest integer less than or equal to $n/2$.    □

**Example 2.7c**  A *partition* of the set $S = \{1, 2, \ldots, n\}$ is a set of mutually exclusive nonempty subsets of $S$ whose union is equal to $S$. That is, we partition $S$ when we split up its elements into mutually exclusive subsets. If $n = 1$ then there is only a single partition, namely $\{1\}$;

if $n = 2$ then there are two possible partitions, $\{\{1, 2\}\}$ and $\{\{1\}, \{2\}\}$. If we let $N(n)$ denote the number of partitions of a set $S$ of size $n$, then we can derive a recursive formula for $N(n)$ as follows.

Focus on one of the elements of $S$, say element $n$, and let us determine the number of partitions of $S$ for which element $n$ is in a subset of size $k$. Noting that there are $\binom{n-1}{k-1}$ different choices of the other elements in its subset and $N(n - k)$ ways of partitioning up the remaining $n - k$ elements (where $N(0)$ is taken to equal 1), it follows that there are $N(n - k)\binom{n-1}{k-1}$ such partitions. Hence, we see that

$$N(n) = \sum_{k=1}^{n} N(n - k)\binom{n-1}{k-1}. \qquad (2.4)$$

Starting with $N(0) = N(1) = 1$, we can recursively use the preceding to calculate $N(2)$, then $N(3)$, and so on. □

**Example 2.7d** In this example we use the recursion equation approach to re-derive the expression given in Example 2.6b for the number of permutations $i_1, i_2, \ldots, i_n$ of $1, 2, \ldots, n$ for which $i_j \neq j$ for each $j = 1, \ldots, n$. Let $N(n)$ denote the number of such permutations (called derangements), and consider the number of them having $i_1 = 2$. Such derangements can have either:

(a) $i_1 = 2$ and $i_2 = 1$; or
(b) $i_1 = 2$ and $i_2 \neq 1$.

The number of derangements of type (a) is equal to the number of permutations $i_3, \ldots, i_n$ of $3, \ldots, n$ for which $i_j \neq j$ ($j = 3, \ldots, n$). Consequently, there are $N(n - 2)$ derangements of type (a). The number of derangements of type (b) is the number of permutations $i_2, i_3, \ldots, i_n$ of $1, 3, \ldots, n$ in which the $j$th element of the permutation is unequal to the $j$th smallest of the values being permuted. Consequently, there are $N(n - 1)$ derangements of type (b). Therefore, there are a total of $N(n-1) + N(n-2)$ derangements that have $i_1 = 2$. As there are clearly an equal number that have $i_1 = j$ for any $j = 3, \ldots, n$, we see that

$$N(n) = (n - 1)N(n - 1) + (n - 1)N(n - 2). \qquad (2.5)$$

Now let $F(n)$ be the fraction of all permutations that are derangements; that is, let

$$F(n) = \frac{N(n)}{n!}.$$

The recursion equation (2.5) can be written as

$$n!\,F(n) = (n-1)(n-1)!\,F(n-1) + (n-1)!\,F(n-2)$$
$$= n!\,F(n-1) - (n-1)!\,F(n-1) + (n-1)!\,F(n-2)$$

or (equivalently) as

$$n!\,[F(n) - F(n-1)] = -(n-1)!\,[F(n-1) - F(n-2)]$$

or

$$F(n) - F(n-1) = -\frac{1}{n}[F(n-1) - F(n-2)].$$

Starting with $F(1) = 0$ and $F(2) = 1/2$, the preceding yields

$$F(3) - F(2) = -\frac{1}{3}[F(2) - F(1)] = -\frac{1}{3!},$$

$$F(4) - F(3) = -\frac{1}{4}[F(3) - F(2)] = \frac{1}{4!},$$

$$F(5) - F(4) = -\frac{1}{5}[F(4) - F(3)] = -\frac{1}{5!},$$

and so on. But this yields that

$$F(2) = 1/2!,$$
$$F(3) = 1/2! - 1/3!,$$
$$F(4) = 1/2! - 1/3! + 1/4!,$$
$$F(5) = 1/2! - 1/3! + 1/4! - 1/5!;$$

it is easy to see (and prove by induction) that, for $n > 1$,

$$F(n) = 1/2! - 1/3! + \cdots + (-1)^n/n!.$$

Therefore,

$$N(n) = n! \sum_{i=2}^{n} \frac{(-1)^i}{i!} = n! \sum_{i=0}^{n} \frac{(-1)^i}{i!},$$

which verifies the result of Example 2.6b.     □

**Example 2.7e**  With $S = \{1, 2, \ldots, n\}$, find the number of different sets $S_1, \ldots, S_k$ such that

$$S_1 \subset S_2 \subset \cdots \subset S_k \subset S.$$

*Solution.* Let $N_k(n)$ be the desired number. Let us relate $N_k(n)$ to $N_k(n-1)$ by noting that any set of $k$ increasing subsets of $S$ can be obtained by choosing a set of $k$ increasing subsets of $\{1, 2, \ldots, n-1\}$, call them $A_1 \subset A_2 \subset \cdots \subset A_k$ and then choosing a number $j$ ($j = 0, \ldots, k$). The $k$ subsets of $S$ are then obtained by adding the element $n$ to the largest $j$ of these subsets. That is, if $j > 0$ then add $n$ to the subsets $A_{k-j+1}, \ldots, A_k$; if $j = 0$, leave the subsets as is. As a result, it follows that

$$N_k(n) = (k+1)N_k(n-1).$$

Starting with $N_k(0) = 1$ (when $S$ has no elements, all $S_i$ must equal the null set), we have

$$N_k(1) = k+1,$$
$$N_k(2) = (k+1)N_k(1) = (k+1)^2,$$
$$N_k(3) = (k+1)N_k(2) = (k+1)^3;$$

it is easy to see that

$$N_k(n) = (k+1)^n.$$

Another way to solve this problem is to consider the number of sequences of sets $S_1 \subset S_2 \subset \cdots \subset S_k \subset S$ for which $S_k$ contains exactly $j$ elements. There are $N_{k-1}(j)$ sequences in which $S_k$ consists of a specified set of $j$ elements as well as $\binom{n}{j}$ different choices of the $j$ elements, so it follows that there are $\binom{n}{j}N_{k-1}(j)$ sequences in which $S_k$ consists of exactly $j$ elements. But this implies that

$$N_k(n) = \sum_{j=0}^{n} \binom{n}{j} N_{k-1}(j).$$

Starting with

$$N_1(j) = 2^j$$

(from Example 2.4f), we obtain

$$N_2(n) = \sum_{j=0}^{n}\binom{n}{j}2^j = (2+1)^n = 3^n,$$

$$N_3(n) = \sum_{j=0}^{n}\binom{n}{j}3^j = (3+1)^n = 4^n.$$

If we now take as the induction hypothesis that $N_{k-1}(j) = k^j$ for each $j = 0, \ldots, n$, it follows that

$$N_k(n) = \sum_{j=0}^{n}\binom{n}{j}k^j = (k+1)^n,$$

which completes the induction proof.  $\square$

**Example 2.7f** *Ballot Problem*  How many different linear orderings are there of $n$ red and $m$ blue balls $(n > m)$ in which there are more red (R) than blue (B) balls among the first $i$ for each $i = 1, \ldots, n+m$, considering same-colored balls to be indistinguishable from each other? (For instance, when $n = 3$ and $m = 2$ there are two orderings: RRRBB and RRBRB.)

*Solution.* Let $N(n, m)$ denote the desired number of orderings. The number of these orderings that have a red ball in the last position is equal to the number of different orderings of $n - 1$ red and $m$ blue balls in which there are always more red than blue balls among the first $i$ for each $i = 1, \ldots, n+m-1$, so it follows that there are $N(n-1, m)$ such orderings. Similarly, there are $N(n, m-1)$ orderings of the desired type that have a blue ball in the last position. Therefore,

$$N(n, m) = N(n-1, m) + N(n, m-1). \qquad (2.6)$$

Starting with $N(1, 0) = 1$ and $N(i, i) = 0$, equation (2.6) can be used recursively to find the values of $N(n, m)$ for specified $n$ and $m$. However, to obtain a general formula it is more useful to work with the quantities $F(n, m)$, defined by

$$F(n, m) = \frac{N(n, m)}{\binom{n+m}{n}}.$$

There are $\binom{n+m}{n}$ different linear orderings of the $n$ red and $m$ blue balls (since there are that number of different choices of the $n$ locations to put the red balls). Hence we see that $F(n, m)$ is the fraction of all orderings that are of the type being considered.

Equation (2.6) can be written as

$$\binom{n+m}{n} F(n, m) = \binom{n+m-1}{n-1} F(n-1, m)$$
$$+ \binom{n+m-1}{n} F(n, m-1)$$

or (equivalently)

$$\frac{(n+m)!}{n!\, m!} F(n, m) = \frac{(n+m-1)!}{(n-1)!\, m!} F(n-1, m)$$
$$+ \frac{(n+m-1)!}{n!\, (m-1)!} F(n, m-1)$$

or

$$F(n, m) = \frac{n}{n+m} F(n-1, m) + \frac{m}{n+m} F(n, m-1). \qquad (2.7)$$

We will now show, by induction on the quantity $k = n + m$, that the solution to (2.7) is

$$F(n, m) = \frac{n-m}{n+m}, \quad n \geq m.$$

To begin, let $S_k$ be the statement that $F(n, m) = (n - m)/(n + m)$ whenever $m \leq n$ and $n + m = k$. Note that the formula is correct when $k = 1$ (since $F(1, 0) = 1$ and $F(1, 1) = 0$). Suppose now that all the statements $S_1, \ldots, S_k$ are true, and suppose that $m \leq n$ and $n + m = k+1$. Clearly $F(n, m) = 0$ when $n = m$, so suppose that $n > m$. Then, by equation (2.7),

$$F(n, m) = \frac{n}{n+m} F(n-1, m) + \frac{m}{n+m} F(n, m-1)$$
$$= \frac{n}{n+m} \frac{n-m-1}{n+m-1} + \frac{m}{n+m} \frac{n-m+1}{n+m-1}$$
$$= \frac{(n+m)(n-m) - n + m}{(n+m)(n+m-1)}$$

$$= \frac{(n-m)(n+m-1)}{(n+m)(n+m-1)}$$

$$= \frac{n-m}{n+m}$$

and the induction proof is complete. (Note that the second equality follows from the induction hypothesis, since $n-1+m = k$.) Therefore, the number of orderings that always have more red than blue balls among the first $i$ (for each $i = 1, \ldots, n+m$) is

$$N(n, m) = \frac{n-m}{n+m} \binom{n+m}{n}.$$

This is known as the "ballot" problem because if we consider an election between two candidates then we can view a red ball as a ballot for candidate A and a blue ball as a ballot for candidate B. Then $N(n, m)$ would represent the number of different counts of the $n+m$ ballots, $n$ for A and $m$ for B, in which A is always leading. □

**Example 2.7g**   Find the number of vectors $(x_1, x_2, x_3)$ where each $x_i$ is an integer and $1 \le x_1 \le x_2 \le x_3 \le 10$.

*Solution.*   Let us generalize the question and let $N_k(n)$ denote the number of vectors $(x_1, \ldots, x_k)$ such that $1 \le x_1 \le x_2 \le \cdots \le x_k \le n$. To obtain a recursion formula for $N_k(n)$, consider the number of such vectors in which $x_k = j$. Since any such vector must satisfy $1 \le x_1 \le x_2 \le \cdots \le x_{k-1} \le j$, it follows that there are $N_{k-1}(j)$ such vectors. Since $x_k$ must equal one of the values $1, \ldots, n$, we see that

$$N_k(n) = \sum_{j=1}^{n} N_{k-1}(j).$$

Starting with $N_1(n) = n$, we obtain

$$N_1(n) = n,$$

$$N_2(n) = \sum_{j=1}^{n} N_1(j)$$

$$= \sum_{j=1}^{n} j$$

$$= n(n+1)/2,$$

$$N_3(n) = \sum_{j=1}^{n} N_2(j)$$

$$= \sum_{j=1}^{n} \frac{j(j+1)}{2}$$

$$= \sum_{j=1}^{n} \frac{j^2}{2} + \sum_{j=1}^{n} \frac{j}{2}$$

$$= \frac{n(n+1)(2n+1)}{12} + \frac{n(n+1)}{4},$$

where the final equality used the identity

$$\sum_{j=1}^{n} j^2 = \frac{n(n+1)(2n+1)}{6}.$$

Thus, the number of possible vectors of the type wanted is $N_3(10) = 220$.  □

## 2.8    The Pigeonhole Principle

The *pigeonhole principle* states that if more than $n$ objects are to be placed in $n$ pigeonholes then at least one pigeonhole will contain more than one object. That this intuitively obvious result can be quite useful is illustrated by the following examples.

**Example 2.8a**   All 82 entering students of a certain high school take courses in English, history, math, and science. If there are three sections of each of these four subjects, show that there are two students that have all four classes together.

*Solution.* Arbitrarily number the different sections of each subject as sections 1, 2, and 3. Now classify a student as being of type $i, j, k, r$ if the student is in English section $i$, history section $j$, math section $k$, and science section $r$. Then, since there are 82 students and only $3^4 = 81$ classifications, it follows from the pigeonhole principle that at least two students will have the same classification.  □

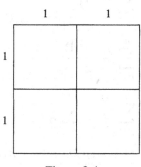

Figure 2.4

**Example 2.8b**    Five points are located inside a square whose sides are of length 2. Show that two of the points are within a distance $\sqrt{2}$ of each other.

**Solution.**    Divide up the square into four square regions of area 1, as indicated in Figure 2.4. By the pigeonhole principle, it follows that at least one of these regions will contain at least two points. The result now follows since two points in a square of radius 1 cannot be further apart then the length of the diagonal of that square, which (by the Pythagorean theorem) is $\sqrt{2}$.    □

The following application of the pigeonhole principle is considered to be classical.

**Example 2.8c**    Show that every sequence of at least $n^2 + 1$ distinct numbers contains either an increasing or a decreasing subsequence of size $n + 1$.

**Solution.**    Let the sequence be $x_1, \ldots, x_i, \ldots, x_N$, where $N \geq n^2 + 1$. Let $a_i$ and $b_i$ be, respectively, the lengths of the longest increasing and decreasing subsequences starting with $x_i$. If any of the $a_i$ or $b_i$ is at least $n + 1$ then we are finished, so suppose that each of these values is one of the numbers $1, 2, \ldots, n$. Consequently, each pair $(a_i, b_i)$ can have $n^2$ possible values. As there are $N \geq n^2 + 1$ pairs, it follows from the pigeonhole principle that at least two of the pairs have the same values. That is, for some $i < j$ we have $a_i = a_j$ and $b_i = b_j$. However, if

$x_i < x_j$ then the subsequence starting with $x_i$ and then continuing with the longest increasing subsequence starting with $x_j$ will be an increasing subsequence that starts with $x_i$ and is of length $1 + a_j = 1 + a_i$, which contradicts our assumption that $a_i$ is the length of the longest increasing subsequence starting at $x_i$. Similarly, if $x_i > x_j$, then the subsequence starting with $x_i$ and continuing with the longest decreasing subsequence beginning with $x_j$ will be an decreasing subsequence that starts with $x_i$ and is of length $1 + b_i$, which is a contradiction. Thus, assuming that there is no monotone subsequence of size $n + 1$ leads to a contradiction, which proves the result. □

A more general (though no less obvious) version of the pigeonhole principle states that if more than $nk$ objects are placed in $n$ pigeonholes, then at least one pigeonhole will contain more than $k$ objects.

**Example 2.8d** Suppose that 14 of the 48 beads of a necklace are colored. Show that there is a string of seven consecutive beads of which at least three are colored.

*Solution.* To begin, note that it is not possible for every string of six consecutive beads to contain at least two that are colored. For since the 48 beads consist of eight nonoverlapping strings of six beads each, this would require at least 16 colored beads. Thus, we can assume that there is a string of six consecutive beads that contains fewer than two colored ones. But this means that the remaining string of 42 beads contains more than 12 colored ones. Consequently, if we divide these 42 beads into six consecutive nonoverlapping strings of seven each, then it follows from the pigeonhole principle that at least one of these nonoverlapping strings contains more than two colored beads. □

## 2.9 Exercises

**Exercise 2.1** How many different seven-place license numbers are possible if the first two places are for letters and the other five are for numbers? What if no letter or number can be repeated?

**Exercise 2.2** Use mathematical induction to prove the generalized basic principle of counting.

**Exercise 2.3**  A simplified model of the stock market assumes that, in each period, the price of a stock can do one of the following: go up 1, go down 1, or remain the same. Under this model, how many different possible results can a stock have over a ten-day period?

**Exercise 2.4**  In how many ways can four boys and four girls sit in a row?

(a)  What if the boys and the girls are each to sit together?
(b)  What if only the boys must sit together?
(c)  What if no two people of the same sex are allowed to sit together?

**Exercise 2.5**  How many different letter arrangements can be made from the letters in:  (a) PEPPER; (b) FLUKE; (c) PROPOSE; (d) MIS-SISSIPPI?

**Exercise 2.6**  A child has 14 blocks, of which six are black, four are yellow, three are green, and one is yellow. If the blocks are put in a row, how many arrangements are possible?

**Exercise 2.7**  If $10! = 3,628,800$, then what is $11!$ equal to?

**Exercise 2.8**  How many permutations of $1, 2, \ldots, 9$ have exactly three digits between 1 and 2?

**Exercise 2.9**  How many permutations are there of $1, 2, \ldots, n$ for which 1 precedes 2 precedes 3?

**Exercise 2.10**  How many permutations of $1, 2, \ldots, 2n$ have every even number in an even-numbered position?

**Exercise 2.11**  How many subsets of $\{1, 2, \ldots, 2n\}$ contain exactly $k$ odd numbers?

**Exercise 2.12**  List how many ways five novels, four mathematics books, and two chemistry books can be linearly arranged on a bookshelf if all the books are distinct and:

(a)  the books can be arranged in any order;
(b)  the mathematics books must be together and the novels must be together;

(c) the novels must be together, but the other books can be arranged in any order.

**Exercise 2.13**    You have twelve different postcards that you want to send to five friends, each of whom can be sent any number of postcards. How many different choices do you have?

**Exercise 2.14**    Use the definition of $\binom{n}{i}$ to show that

$$\binom{n}{j} = \binom{n}{n-j};$$

then give a combinatorial explanation for this identity.

**Exercise 2.15**    If $p$ is prime, show that it divides $\binom{p}{i}$ for every $i = 1, \ldots, p-1$.

**Exercise 2.16**    A dance class consists of 22 students, 10 women and 12 men. If five men and five women are to be chosen and then paired off, how may results are possible?

**Exercise 2.17**    If $2n$ people are to be divided into $n$ pairs of two each, how many different divisions are possible?

**Exercise 2.18**    A student must sell two books from a collection of six mathematics, seven science, and four economics books. Show how many choices are possible if

(a) both books must be on the same subject;
(b) the books are to be on different subjects.

**Exercise 2.19**    Give a combinatorial explanation of the fact that there are $\binom{n}{k}$ different linear arrangements of $n$ balls of which $k$ are black and $n-k$ are white.

**Exercise 2.20**    Determine the number of vectors $(x_1, \ldots, x_n)$ such that (i) each $x_i$ is either 0 or 1 and (ii)

$$\sum_{i=1}^{n} x_i \geq k.$$

**Exercise 2.21**    How many vectors $(x_1, \ldots, x_k)$ are there for which each $x_i$ is a positive integer such that $1 \leq x_i \leq n$ and $x_1 < x_2 < \cdots < x_k$?

**Exercise 2.22**   Give a combinatorial argument for the identity

$$\binom{n+m}{r} = \binom{n}{0}\binom{m}{r} + \binom{n}{1}\binom{m}{r-1} + \cdots + \binom{n}{r}\binom{m}{0}.$$

*Hint:* Imagine that a group of size $r$ must be chosen from a set of $n$ men and $m$ women.

**Exercise 2.23**   A committee of size $k$, with one of the committee members designated as chairperson, is to be chosen from a set of $n$ people.

(a) By focusing first on the choice of the committee and then on the choice of the chair, argue that there are $\binom{n}{k}k$ possible choices.
(b) Focusing first on nonchair committee members and then on the chairperson, argue that there are $\binom{n}{k-1}(n-k+1)$ possible choices.
(c) Focusing first on the choice of the chair and then on the choice of the other committee members, argue that there are $n\binom{n-1}{k-1}$ possible choices.
(d) Conclude from the preceding that

$$k\binom{n}{k} = (n-k+1)\binom{n}{k-1} = n\binom{n-1}{k-1}.$$

(e) Use the factorial definition of $\binom{n}{r}$ to verify the identity in part (d).

**Exercise 2.24**   The following expression is known as Fermat's combinatorial identity:

$$\binom{n}{k} = \sum_{i=k}^{n}\binom{i-1}{k-1}.$$

Give a combinatorial argument for this identity.
   *Hint:* Consider the set of numbers from 1 to $n$. How many subsets of size $k$ have $i$ as their highest-numbered member?

**Exercise 2.25**   An elevator starts at the basement with eight passengers (not including the elevator operator) and discharges them all by the time it reaches the top (sixth) floor. In how many ways could the operator have perceived the people leaving the elevator if all people look alike to him? What if the eight people consisted of five women and three men (assuming the operator could distinguish a woman from a man)?

**Exercise 2.26**  Determine the number of vectors $(x_1, \ldots, x_n)$ such that each $x_i$ is a nonnegative integer and

$$\sum_{i=1}^{n} x_i \le k.$$

**Exercise 2.27**  A total of \$20,000 must be invested among four possible opportunities. Each investment must be in integral units of \$1 thousand, and the minimal investments for the opportunities are \$2, \$2, \$3, and \$4 thousand, respectively. State how many different investment strategies are possible if:

(a)  an investment must be made in each opportunity;
(b)  investments must be made in at least three of the four opportunities.

**Exercise 2.28**  All thirteen staff members in a United Nations office in New York know at least one of the languages French, Spanish, or German. Ten of them know Spanish, seven know German, six know French, five know both Spanish and German, four know both Spanish and French, and three know both French and German.

(a)  How many know all three languages?
(b)  How many know exactly two of the three languages?
(c)  How many know exactly one of the three languages?

**Exercise 2.29**  A system is composed of five components, each of which is either working or failed. Consider an experiment that consists of observing the status of each component, and let the outcome of the experiment be given by the vector $(x_1, x_2, x_3, x_4, x_5)$, where $x_i$ is equal to 1 if component $i$ is working or equal to 0 if component $i$ failed.

(a)  How many outcomes are possible?
(b)  Suppose that the system will work if components 1 and 2 are both working, or if components 2 and 3 are both working, or if components 1, 3, and 5 are all working. Let $W$ be the set of all outcomes for which the system works. How many outcomes are in $W$?

**Exercise 2.30**  Find the number of positive integers less than or equal to 1,000 that are divisible by 3, 5, or 7.

**Exercise 2.31**    In Example 2.6c, let $N(n, m)$ denote the number of outcomes in which at least one coupon of each type is collected. Derive the recursion equation

$$N(n, m) = \sum_{j=1}^{m} \binom{m}{j} N(n - 1, m - j).$$

**Exercise 2.32**    In Example 2.7b, show by mathematical induction that

$$N(n) = 2^{2n-1} + 2^{n-1}.$$

**Exercise 2.33**    Use the recursion equation (2.4) to derive the number of partitions of a set consisting of four elements.

**Exercise 2.34**    Derive a recursion equation for the number of strings of 0s and 1s of length $n$ in which the substring 0, 0 never appears. Use your equation to find the desired number when $n = 8$.

   *Hint:* How many such strings begin with 0? How many with 1?

**Exercise 2.35**    Derive a recursion for the number of subsets of the set $\{1, 2, \ldots, n\}$ that do not contain consecutive integers, and then use it to find the answer when $n = 7$.

   *Hint:* How many such subsets have $j$ as their largest element?

**Exercise 2.36**    Consider a tournament of $n$ contestants in which the outcome of the tournament is an ordering of these contestants (ties are allowed). That is, the outcome partitions the players into groups, with the first group consisting of the players that tied for first place, the next group being those that tied for the next best position, and so on. Let $N(n)$ denote the number of different possible outcomes. For instance, $N(2) = 3$ because – in a tournament with two contestants – each player could be uniquely first or the players could tie for first place.

(a)  List all possible outcomes when $n = 3$.

(b)  With $N(0)$ defined to equal 1, argue (without any computations) that

$$N(n) = \sum_{i=1}^{n} \binom{n}{i} N(n - i).$$

   *Hint:* How many outcomes are there in which $i$ players tie for last place?

(c)  Show that the formula of part (b) is equivalent to

$$N(n) = \sum_{i=1}^{n-1} \binom{n}{i} N(i).$$

(d)  Use the preceding to find $N(3)$ and $N(4)$.

**Exercise 2.37**   For any permutation $i_1, i_2, \ldots, i_n$ of $1, 2, \ldots, n$, we say that the ordered pair $(i, j)$ is an *inversion* of the permutation if $i < j$ and $j$ precedes $i$ in the permutation. Let $N_k(n)$ denote the number of permutations of $1, 2, \ldots, n$ that have exactly $k$ inversions. Argue that

$$N_k(n + 1) = N_k(n) + N_{k-1}(n + 1).$$

**Exercise 2.38**   Say that an $n$-digit sequence of 0s and 1s is acceptable if it has the property that each 0 in the sequence is next to at least one other 0, and let $A(n)$ denote the number of such sequences. Thus $A(1) = 1$, since the only acceptable one-digit sequence is 1; and $A(2) = 2$, since the acceptable two-digit sequences are 1, 1 and 0, 0.

(a)  Find $A(3)$.
(b)  With $A(0)$ defined to equal 1, show that

$$A(n) = 1 + A(n - 1) + \sum_{i=3}^{n} A(n - i)$$

for $n \geq 3$.
   *Hint*: How many acceptable sequences have their first 1 appearing in position $i$?
(c)  Show that, for $n \geq 3$,

$$A(n + 1) = 2A(n) - A(n - 1) + A(n - 2).$$

(d)  Find $A(8)$.

**Exercise 2.39**   From a set of 121 female mice, show that there must either be (i) a set of twelve mice consisting of daughter, mother, grandmother, great-grandmother, ... or (ii) a set of eleven mice no one of which is the mother of any other.

**Exercise 2.40**   If 46 pigeons are put into 10 pigeonholes, show that, for some $i$, there are at least $i$ pigeons in pigeonhole number $i$.

# 3. Probability

## 3.1 Probabilities and Events

Consider an experiment and let $S$, called the *sample space,* be the set of all possible outcomes. If there are $m$ possible outcomes of the experiment, then we will generally number them 1 through $m$, and so $S = \{1, 2, \ldots, m\}$. (However, when dealing with specific examples, we will usually give more descriptive names to the outcomes.)

**Example 3.1a** (i) Let the experiment consist of the flipping of a coin, and let the outcome be the side that lands face up. Thus, the sample space of this experiment is

$$S = \{h, t\},$$

where the outcome is $h$ if the coin shows heads and $t$ if it shows tails.

(ii) If the experiment consists of rolling a pair of dice and the outcome is the pair $(i, j)$, where $i$ is the value that appears on the first die and $j$ the value on the second, then the sample space consists of the following 36 outcomes:

$$(1, 1), \ (1, 2), \ (1, 3), \ (1, 4), \ (1, 5), \ (1, 6),$$
$$(2, 1), \ (2, 2), \ (2, 3), \ (2, 4), \ (2, 5), \ (2, 6),$$
$$(3, 1), \ (3, 2), \ (3, 3), \ (3, 4), \ (3, 5), \ (3, 6),$$
$$(4, 1), \ (4, 2), \ (4, 3), \ (4, 4), \ (4, 5), \ (4, 6),$$
$$(5, 1), \ (5, 2), \ (5, 3), \ (5, 4), \ (5, 5), \ (5, 6),$$
$$(6, 1), \ (6, 2), \ (6, 3), \ (6, 4), \ (6, 5), \ (6, 6).$$

(iii) If the experiment consists of a horse race of $r$ horses (numbered 1, 2, 3, $\ldots$, $r$) and the outcome is the order of finish of these horses, then the sample space is

$S = \{\text{all orderings of the numbers } 1, 2, 3, \ldots, r\}.$

For instance, if $r = 4$, then the outcome is $(1, 4, 2, 3)$ if the number-1 horse comes in first, number 4 comes in second, number 2 comes in third, and number 3 comes in fourth. □

Consider again an experiment with the sample space $S = \{1, 2, \ldots, m\}$. We will now suppose that there are numbers $p_1, \ldots, p_m$ with

$$p_i \geq 0, \quad i = 1, \ldots, m, \quad \text{and} \quad \sum_{i=1}^{m} p_i = 1$$

and such that $p_i$ is the *probability* that $i$ is the outcome of the experiment.

**Example 3.1b** In Example 3.1a(i), the coin is said to be *fair* or *unbiased* if it is equally likely to land on heads as on tails. Thus, for a fair coin we would have that

$$p_h = p_t = 1/2.$$

If the coin were biased such that heads were twice as likely to appear as tails, then we would have

$$p_h = 2/3, \qquad p_t = 1/3.$$

If an unbiased pair of dice were rolled in Example 3.1a(ii), then all possible outcomes would be equally likely and so

$$p_{(i,j)} = 1/36, \qquad 1 \leq i \leq 6, \quad 1 \leq j \leq 6.$$

If $r = 3$ in Example 3.1a(iii), then we suppose that we are given the six nonnegative numbers that sum to 1:

$$p_{1,2,3}, \ p_{1,3,2}, \ p_{2,1,3}, \ p_{2,3,1}, \ p_{3,1,2}, \ p_{3,2,1},$$

where $p_{i,j,k}$ represents the probability that horse $i$ comes in first, horse $j$ second, and horse $k$ third. □

Any set of possible outcomes of the experiment is called an *event*. That is, an event is a subset of $S$, the set of all possible outcomes. For any event $A$, we say that $A$ occurs whenever the outcome of the experiment is a point in $A$. If we let $P(A)$ denote the probability that the event $A$ occurs, then we can determine it by using the equation

$$P(A) = \sum_{i \in A} p_i. \tag{3.1}$$

Note that this implies

$$P(S) = \sum_i p_i = 1. \tag{3.2}$$

That is, the probability that the outcome of the experiment is in the sample space is equal to 1, which – since $S$ consists of all possible outcomes of the experiment – is the desired result.

**Example 3.1c**   Suppose that the experiment consists of rolling a pair of fair dice. If $A$ is the event that the sum of the dice is equal to 7, then

$$A = \{(1, 6), (2, 5), (3, 4), (4, 3), (5, 2), (6, 1)\}$$

and

$$P(A) = 6/36 = 1/6.$$

If we let $B$ be the event that the sum is 8, then

$$P(B) = p_{(2,6)} + p_{(3,5)} + p_{(4,4)} + p_{(5,3)} + p_{(6,2)} = 5/36.$$

In a horse race between three horses, if we let $A$ denote the event that horse number 1 wins then $A = \{(1, 2, 3), (1, 3, 2)\}$ and

$$P(A) = p_{1,2,3} + p_{1,3,2}. \qquad \square$$

For any event $A$, let $A^c$, called the *complement* of $A$, be the event containing all those outcomes in $S$ that are not in $A$. That is, $A^c$ occurs if and only if $A$ does not. Since

$$1 = \sum_i p_i$$

$$= \sum_{i \in A} p_i + \sum_{i \notin A} p_i$$

$$= P(A) + P(A^c),$$

we see that

$$P(A^c) = 1 - P(A). \tag{3.3}$$

That is, the probability that the outcome is *not* in $A$ is 1 minus the probability that it *is* in $A$. The complement of the sample space $S$ is the null event $\emptyset$, which contains no outcomes. Since $\emptyset = S^c$, we obtain from equations (3.2) and (3.3) that

$$P(\emptyset) = 0.$$

Events are just subsets of the sample space $S$, so if $A$ and $B$ are events then their *union* $A \cup B$ is the event consisting of all outcomes that are either in $A$ or in $B$. Their *intersection* $AB$ (sometimes written $A \cap B$) is the event consisting of all outcomes that are both in $A$ and in $B$.

**Example 3.1d**   Let the experiment consist of rolling a pair of dice. If $A$ is the event that the sum is 10 and $B$ is the event that both dice land on even numbers greater than 3, then

$$A = \{(4, 6), (5, 5), (6, 4)\} \quad \text{and} \quad B = \{(4, 4), (4, 6), (6, 4), (6, 6)\}.$$

Therefore,

$$A \cup B = \{(4, 4), (4, 6), (5, 5), (6, 4), (6, 6)\},$$
$$AB = \{(4, 6), (6, 4)\}. \qquad \square$$

For any events $A$ and $B$, we can write

$$P(A \cup B) = \sum_{i \in A \cup B} p_i,$$

$$P(A) = \sum_{i \in A} p_i,$$

$$P(B) = \sum_{i \in B} p_i.$$

Since every outcome in both $A$ and $B$ is counted twice in $P(A) + P(B)$ and only once in $P(A \cup B)$, we obtain the following result, often called the *addition theorem of probability*.

**Proposition 3.1.1**

$$P(A \cup B) = P(A) + P(B) - P(AB).$$

Thus, the probability that the outcome of the experiment is either in $A$ or in $B$ is equal to the probability that it is in $A$ plus the probability that it is in $B$ minus the probability that it is in both $A$ and $B$.

If $AB = \emptyset$, we say that $A$ and $B$ are *mutually exclusive* or *disjoint*. That is, events are mutually exclusive if they cannot both occur. Since $P(\emptyset) = 0$, it follows from Proposition 3.1.1 that, when $A$ and $B$ are mutually exclusive,

$$P(A \cup B) = P(A) + P(B).$$

## 3.2    Probability Experiments Having Equally Likely Outcomes

For many experiments it is natural to assume that all outcomes in the sample space are equally likely to occur. That is, if the sample space is $S = \{1, \ldots, m\}$ then it is often natural to suppose that $p_i = 1/m$ for $i = 1, \ldots, m$. As a result, for any event $A$ we have

$$P(A) = \sum_{i \in A} p_i = \frac{|A|}{m},$$

where $|A|$ is the number of outcomes in $A$. In other words, if each outcome is equally likely, then the probability that the outcome is in $A$ is equal to the ratio of the number of outcomes of $S$ that are in $A$ to the number of outcomes in $S$.

**Example 3.2a**    If a pair of fair dice is rolled, what is the probability that the sum of the dice is one of the values 2, 3, or 12?

*Solution.*    If $A$ is the event that the sum is 2, 3, or 12, then

$$A = \{(1, 1), (1, 2), (2, 1), (6, 6)\}.$$

Therefore,

$$P(A) = 4/36 = 1/9. \qquad \square$$

**Example 3.2b** A five-card poker hand is said to be a "full house" if three of its cards are the same denomination and the other two are the same denomination. (That is, a full house is three of a kind plus a pair.) If you are dealt five cards from a well-shuffled deck of 52 playing cards, what is the probability that you are dealt a full house?

*Solution.* Let us assume that, in a well-shuffled deck, all $\binom{52}{5}$ hands are equally likely. To determine the number of these hands that are full houses, note first that there 13 possible choices for the denomination of the three of a kind and, for a given choice of denomination, there are $\binom{4}{3}$ choices of the three cards of that denomination. Also, given the choice of the denomination of the trio, there are 12 possible choices for the denomination of the pair, and then $\binom{4}{2}$ choices of the two cards. Thus, we see that

$$P(\text{full house}) = \frac{13 \cdot \binom{4}{3} \cdot 12 \cdot \binom{4}{2}}{\binom{52}{5}} \approx 0.0014. \qquad \square$$

Our next example illustrates that probability results can be quite surprising when initially encountered.

**Example 3.2c** If $n$ people are present in a room, what is the probability that no two of them celebrate their birthday on the same day of the year? How large need $n$ be so that this probability is less than $1/2$?

*Solution.* Each person can celebrate his or her birthday on any one of 365 days, so there are $(365)^n$ possible outcomes. (We assume here that nobody was born on February 29.) The number of these outcomes that result in no identical birthdays is $365 \cdot 364 \cdot 363 \cdots (365 - (n-1))$. This can be seen by noting that there will be no matches if the first birthday is any of the 365 days, the next birthday is then any of the remaining 364 days, the next any of the remaining 363 days, and so on. Assuming that each outcome is equally likely, we have that

$$P(\text{no birthday matches}) = \frac{365 \cdot 364 \cdot 363 \cdots (365 - (n-1))}{365^n}.$$

People are often surprised to hear that this probability is less than $1/2$ when $n = 23$. That is, if there are 23 (or more) people in a room then the probability that at least two of them share the same birthday exceeds 1/2. This appears surprising, since 23 seems small in comparison with 365. However, every pair of individuals has probability $365/(365)^2 = 1/365$ of sharing the same birthday, and in a group of 23 people there are $\binom{23}{2} = 253$ distinct pairs. Looked at this way, the result no longer seems surprising. □

**Example 3.2d**    If three balls are randomly drawn from a bowl that contains seven white and five black balls, what is the probability that one of the drawn balls is white and the other two black?

*Solution.*  If we regard the order in which the balls are selected as being relevant, then the sample space consists of $12 \cdot 11 \cdot 10 = 1{,}320$ distinct outcomes. Furthermore, there are $7 \cdot 5 \cdot 4 = 140$ outcomes in which the first ball selected is white and the other two black, as well as the same number of outcomes for which the second ball is white and the others black, and the same number for which the third is white and the others black. Hence, assuming that "randomly drawn" means that each of the outcomes in the sample space is equally likely to occur, the desired probability is

$$\frac{3 \cdot 140}{1{,}320} = \frac{7}{22}.$$

This problem could also have been solved by regarding the outcome of the experiment as the unordered set of balls drawn. From this point of view, there are $\binom{12}{3}$ outcomes in the sample space, with $\binom{7}{1}\binom{5}{2}$ of these outcomes resulting in the selection of one white and two black balls. Now each set of three balls corresponds to 3! outcomes when the order of selection is noted. Hence, if all outcomes are assumed to be equally likely when the order of selection is taken into account, they should remain equally likely when the order is no longer considered relevant. Using this latter representation of the experiment thus shows that the desired probability is

$$\frac{\binom{7}{1}\binom{5}{2}}{\binom{12}{3}} = \frac{7}{22},$$

which (of course) agrees with the previously obtained answer. □

**Example 3.2e**   A committee of four is to be selected from a group of six men and twelve women. If the selection is random, what is the probability that the committee consists of two men and two women?

*Solution.*   Assuming that the selection being "random" means that each of the $\binom{18}{4}$ possible committees is equally likely to be selected, the desired probability is

$$\frac{\binom{6}{2}\binom{12}{2}}{\binom{18}{4}} = \frac{11}{34}.$$   □

## 3.3    Conditional Probability

Suppose that each of two teams is to produce an item, and that the two items produced will be rated as either acceptable or unacceptable. The sample space of this experiment will then consist of the following four outcomes:

$$S = \{(a, a), (a, u), (u, a), (u, u)\},$$

where, for example, $(a, u)$ means that the first team produced an acceptable item and the second team an unacceptable one. Suppose that the probabilities of these outcomes are as follows:

$$P(a, a) = 0.54,$$
$$P(a, u) = 0.28,$$
$$P(u, a) = 0.14,$$
$$P(u, u) = 0.04.$$

If we are given the information that exactly one of the items produced was acceptable, what is the probability that it was the one produced by the first team? To determine this probability, consider the following reasoning. Given that there was exactly one acceptable item produced, it follows that the outcome of the experiment was either $(a, u)$ or $(u, a)$. Since the outcome $(a, u)$ was initially twice as likely as the outcome $(u, a)$, it should remain twice as likely given the information that one of them occurred. Therefore, the probability that the outcome was $(a, u)$ is $2/3$, whereas the probability that it was $(u, a)$ is $1/3$.

Let $A = \{(a, u), (a, a)\}$ denote the event that the item produced by the first team is acceptable, and let $B = \{(a, u), (u, a)\}$ be the event that exactly one of the produced items is acceptable. The probability that the item produced by the first team was acceptable given that exactly one of the produced items was acceptable is called the *conditional probability* of $A$ given that $B$ has occurred, and is denoted as

$$P(A|B)$$

A general formula for $P(A|B)$ is obtained by an argument similar to the one just given. Namely, if the event $B$ occurs, then in order for the event $A$ to occur it is necessary that the occurrence be a point in both $A$ and $B$; that is, it must be in $AB$. Now, since we know that $B$ has occurred, it follows that $B$ can be thought of as the new sample space, and hence the probability that the event $AB$ occurs will equal the probability of $AB$ relative to the probability of $B$. That is,

$$P(A|B) = \frac{P(AB)}{P(B)}. \tag{3.4}$$

**Example 3.3a**   A coin is flipped twice. Assuming that all four points in the sample space $S = \{(h, h), (h, t), (t, h), (t, t)\}$ are equally likely, what is the conditional probability that both flips land on heads, given that:

(a)  the first flip lands on heads;
(b)  at least one of the flips lands on heads?

**Solution.** Let $A = \{(h, h)\}$ be the event that both flips land on heads; let $B = \{(h, h), (h, t)\}$ be the event that the first flip lands on heads; and let $C = \{(h, h), (h, t), (t, h)\}$ be the event that at least one of the flips lands on heads. We have the following solutions:

$$P(A|B) = \frac{P(AB)}{P(B)}$$
$$= \frac{P(\{(h, h)\})}{P(\{(h, h), (h, t)\})}$$
$$= \frac{1/4}{2/4}$$
$$= 1/2$$

and

$$P(A|C) = \frac{P(AC)}{P(C)}$$

$$= \frac{P(\{(h, h)\})}{P(\{(h, h), (h, t), (t, h)\})}$$

$$= \frac{1/4}{3/4}$$

$$= 1/3$$

Many people are initially surprised that the answers to parts (a) and (b) are not identical. To understand why the answers are different, note first that – conditional on the first flip landing on heads – the second one is still equally likely to land on either heads or tails and so the probability in part (a) is $1/2$. On the other hand, knowing that at least one of the flips lands on heads is equivalent to knowing that the outcome is not $(t, t)$. Thus, given that at least one of the flips lands on heads, there remain three equally likely possibilities – namely $(h, h)$, $(h, t)$, and $(t, h)$. This shows that the answer to part (b) is $1/3$. □

It follows from equation (3.1) that

$$P(AB) = P(B)P(A|B). \tag{3.5}$$

That is, the probability that both $A$ and $B$ occur is the probability that $B$ occurs multiplied by the conditional probability that $A$ occurs given that $B$ occurred; this result is often called the *multiplication theorem of probability*.

**Example 3.3b** Suppose that two balls are to be withdrawn, without replacement, from an urn that contains nine blue and seven yellow balls. If each ball withdrawn is equally likely to be any of the balls in the urn at the time, what is the probability that both withdrawn balls are blue?

*Solution.* Let $B_1$ and $B_2$ denote, respectively, the events that the first and second balls withdrawn are blue. Now, given that the first ball withdrawn is blue, the second ball is equally likely to be any of the remaining 15 balls, of which 8 are blue. Therefore, $P(B_2|B_1) = 8/15$. As $P(B_1) = 9/16$, we see that

$$P(B_1 B_2) = \frac{9}{16}\frac{8}{15} = \frac{3}{10}.$$

The preceding could also have been obtained as follows:

$$P(B_1 B_2) = \frac{\binom{9}{2}}{\binom{16}{2}} = \frac{3}{10}. \qquad \square$$

The conditional probability of $A$, given that $B$ has occurred, is not generally equal to the unconditional probability of $A$. In other words, knowing that the outcome of the experiment is an element of $B$ generally changes the probability that it is an element of $A$. (What if $A$ and $B$ are mutually exclusive?) In the special case where $P(A|B)$ is equal to $P(A)$, we say that $A$ is *independent* of $B$. Because

$$P(A|B) = \frac{P(AB)}{P(B)},$$

we see that $A$ is independent of $B$ if

$$P(AB) = P(A)P(B). \tag{3.6}$$

This relation is symmetric in $A$ and $B$, so it follows that, whenever $A$ is independent of $B$, $B$ is also independent of $A$ – that is, $A$ and $B$ are *independent events*.

**Example 3.3c**   Suppose that each item produced by a firm is, independently of the quality of other items, of acceptable quality with probability 0.99. Find the probability that two successively produced items are both of acceptable quality.

*Solution.* Let $A_i$ be the event that item $i$ is of acceptable quality. Then, by independence,

$$P(A_1 A_2) = P(A_1)P(A_2) = 0.99^2 = 0.9801. \qquad \square$$

## 3.4    Computing Probabilities by Conditioning

Suppose that $B_1, B_2, \ldots, B_n$ are mutually exclusive events whose union is the sample space $S$. That is, exactly one of these events must occur. Then, for any event $A$, we can write

$$A = \bigcup_i AB_i.$$

Since the events $AB_i$ are all mutually exclusive, the preceding yields that

$$P(A) = \sum_{i=1}^{n} P(AB_i)$$

$$= \sum_{i=1}^{n} P(A|B_i)P(B_i). \qquad (3.7)$$

Equation (3.7) shows that $P(A)$, the probability of event $A$, is a weighted average of the conditional probabilities of $A$ given $B_i$ for $i = 1, \ldots, n$, with $P(A|B_i)$ weighted by the probability that $B_i$ occurs. It shows how we can compute the probability of $A$ by first "conditioning" on which of the events $B_i$ occurs.

**Example 3.4a**     A bin contains three different types of disposable flashlights. The probability that a type-1 flashlight will give over 100 hours of use is 0.7; the corresponding probabilities for type-2 and type-3 flashlights are 0.4 and 0.3, respectively. If 20% of the flashlights in the bin are type-1, 30% are type-2, and 50% are type-3, what is the probability that a randomly chosen flashlight will give more than 100 hours of use?

*Solution.* Let $A$ denote the event that the chosen flashlight will give over 100 hours of use, and let $B_i$ be the event that a type-$i$ flashlight is chosen, $i = 1, 2, 3$. To compute $P(A)$, we condition on the type of the chosen flashlight and so obtain

$$P(A) = P(A|B_1)P(B_1) + P(A|B_2)P(B_2) + P(A|B_3)P(B_3)$$

$$= (0.7)(0.2) + (0.4)(0.3) + (0.3)(0.5) = 0.41.$$

There is a 41% chance that the flashlight will last for over 100 hours.
□

**Example 3.4b**     Consider the following game played with an ordinary deck of 52 playing cards. The cards are shuffled and then turned over one at a time. At any time, the player can guess that the next card to be turned over will be the ace of spades; if it is, then the player wins. In addition, the player is said to win if the ace of spades has not yet appeared when only one card remains and no guess has yet been made. What is a good strategy, and what is a bad strategy?

***Solution.*** Every strategy has probability 1/52 of winning! To show this, we will use induction to prove the stronger result that, for an $n$-card deck that contains the ace of spades, the probability of winning is $1/n$ no matter what strategy is employed. Since this is clearly true for $n = 1$, assume it to be true for an $(n - 1)$-card deck, and now consider an $n$-card deck. Fix any strategy, and let $p$ denote the probability that this strategy guesses that the first card is the ace of spades. Given that it does, then the player's probability of winning is $1/n$. On the other hand, if the strategy does not guess that the first card is the ace of spades, then the probability that the player wins is the probability that the first card is not the ace of spades (i.e., $(n - 1)/n$) multiplied by the conditional probability of winning given that the first card is not the ace of spades. But this latter conditional probability is equal to the probability of winning when using an $(n - 1)$-card deck containing a single ace of spades; it is thus, by the induction hypothesis, $1/(n - 1)$. Hence, given that the strategy does not guess the first card, the probability of winning is

$$\frac{n - 1}{n} \frac{1}{n - 1} = \frac{1}{n}.$$

Thus, conditioning on whether the first card is guessed shows that

$$P\{\text{win}\} = \frac{1}{n}p + \frac{1}{n}(1 - p) = \frac{1}{n},$$

which completes the induction.     □

**Example 3.4c** *Best Prize Problem*   Suppose that we are to be presented with $n$ distinct prizes in sequence. After being presented with a prize, we must immediately decide whether to accept it or to reject it and consider the next prize. The only information we are given when deciding whether to accept a prize is the relative rank of that prize compared to ones already seen. For instance, when the fifth prize is presented, we learn how it ranks among the five prizes aready seen. Suppose that once a prize is rejected it is lost, and that our objective is to maximize the probability of obtaining the best prize. Assuming that all $n!$ possible orderings of the prizes are equally likely, how well can we do?

*Solution.* Rather surprisingly, we can do quite well. To see this, fix a value $k$ ($0 \leq k < n$) and consider the strategy that rejects the first $k$ prizes and then accepts the first prize to appear thereafter that is better than each of the first $k$. Let $W$ be the event that the best prize is obtained (and so the player wins) when this strategy is employed. Also, for $i = 1, \ldots, n$, let $B_i$ be the event that the best prize is the one in position $i$. Exactly one of the events $B_i$ ($i = 1, \ldots, n$) must occur and so, upon conditioning on which event does occur, we obtain

$$P(W) = \sum_{i=1}^{n} P(W|B_i)P(B_i)$$

$$= \frac{1}{n} \sum_{i=1}^{n} P(W|B_i). \tag{3.8}$$

The final equality follows because the best prize is equally likely to be in any of the $n$ positions and so $P(B_i) = 1/n$ for all $i$. Now, since we are employing the strategy of rejecting the first $k$ prizes, it follows that there is no chance of obtaining the best prize if it is among the first $k$. Consequently,

$$P(W|B_i) = 0 \quad \text{if } i \leq k. \tag{3.9}$$

On the other hand, if the best prize is in position $i$, where $i > k$, then the best prize will be selected if the best of the first $i - 1$ prizes is among the first $k$ (for then none of the prizes in positions $k + 1, k + 2, \ldots, i - 1$ would be selected). However, conditional on the best prize being in position $i$, it is easy to see that all possible orderings of the other prizes remain equally likely, which implies that each of the first $i - 1$ prizes is equally likely to be the best of that batch. Hence, we obtain that

$$P(W|B_i) = \frac{k}{i - 1} \quad \text{if } i > k. \tag{3.10}$$

Equations (3.8), (3.9), and (3.10) then yield

$$P(W) = \frac{k}{n} \sum_{i=k+1}^{n} \frac{1}{i - 1}$$

$$= \frac{k}{n} \sum_{j=k}^{n-1} \frac{1}{j}$$

$$= \frac{k}{n} \left[ \sum_{j=1}^{n-1} \frac{1}{j} - \sum_{j=1}^{k-1} \frac{1}{j} \right]$$

$$\approx \frac{k}{n} [\log(n-1) - \log(k-1)]$$

$$= \frac{k}{n} \log\left( \frac{n-1}{k-1} \right)$$

$$\approx \frac{k}{n} \log\left( \frac{n}{k} \right),$$

where we have used the approximation

$$\sum_{j=1}^{n} \frac{1}{j} \approx \log(n).$$

But we see from the preceding that, if we choose $k$ so that

$$\frac{k}{n} \approx \frac{1}{e},$$

then

$$P(W) \approx \frac{1}{e}.$$

That is, the strategy that lets approximately the fraction $1/e$ of all prizes go by, and then accepts the first one thereafter that is the best yet seen, has probability approximately equal to $1/e \approx 0.37$ of obtaining the best prize.    □

Suppose again that the events $B_i$ $(i \geq 1)$ are mutually exclusive events whose union is the sample space. Suppose that an event $A$ occurs and we want to determine which of the $B_j$ also occurred. We have

$$P(B_j|A) = \frac{P(AB_j)}{P(A)}$$

$$= \frac{P(A|B_j)P(B_j)}{\sum_i P(A|B_i)P(B_i)}. \qquad (3.11)$$

Equation (3.11) is known as *Bayes' formula.*

**Example 3.4d**   Suppose in Example 3.4a that the flashlight lasted over 100 hours. What is the conditional probability that it was a type-$j$ flashlight for $j = 1, 2, 3$?

*Solution.*   The probability is obtained by using Bayes' formula. Using the notation of Example 3.4a, we have

$$P(B_j|A) = \frac{P(AB_j)}{P(A)}$$

$$= \frac{P(A|B_j)P(B_j)}{0.41}.$$

Thus,

$$P(B_1|A) = (0.7)(0.2)/0.41 = 14/41,$$

$$P(B_2|A) = (0.4)(0.3)/0.41 = 12/41,$$

$$P(B_3|A) = (0.3)(0.5)/0.41 = 15/41.$$

For instance, whereas the initial probability that a type-1 flashlight is chosen is only 0.2, the information that the chosen flashlight has lasted over 100 hours raises the probability of this event to $14/41 \approx 0.341$.   □

## 3.5    Random Variables and Expected Values

Numerical quantities whose values are determined by the outcome of the experiment are called *random variables*. For instance, the sum obtained when rolling dice, or the number of heads that result from a series of coin flips, are random variables. Since the value of a random variable is determined by the outcome of the experiment, we can assign probabilities to each of its possible values.

**Example 3.5a**   Let the random variable $X$ denote the sum when a pair of fair dice are rolled. The possible values of $X$ are $2, 3, \ldots, 12$, and they have the following probabilities:

$$P\{X = 2\} = P\{(1, 1)\} = 1/36,$$

$$P\{X = 3\} = P\{(1, 2), (2, 1)\} = 2/36,$$

$$P\{X = 4\} = P\{(1, 3), (2, 2), (3, 1)\} = 3/36,$$

$$P\{X = 5\} = P\{(1, 4), (2, 3), (3, 2), (4, 1)\} = 4/36,$$

$$P\{X = 6\} = P\{(1, 5), (2, 4), (3, 3), (4, 2), (5, 1)\} = 5/36,$$

$$P\{X = 7\} = P\{(1, 6), (2, 5), (3, 4), (4, 3), (5, 2), (6, 1)\} = 6/36,$$

$$P\{X = 8\} = P\{(2, 6), (3, 5), (4, 4), (5, 3), (6, 2)\} = 5/36,$$

$$P\{X = 9\} = P\{(3, 6), (4, 5), (5, 4), (6, 3)\} = 4/36,$$

$$P\{X = 10\} = P\{(4, 6), (5, 5), (6, 4)\} = 3/36,$$

$$P\{X = 11\} = P\{(5, 6), (6, 5)\} = 2/36,$$

$$P\{X = 12\} = P\{(6, 6)\} = 1/36. \qquad \square$$

**Example 3.5b** Consider $n$ flips of a coin, and suppose that the successive flips are independent and that each flip results in heads with probability $p$. The random variable $X$, equal to the total number of heads that occur, is called a *binomial* random variable with parameters $n$ and $p$. To determine the probabilities attached to the possible values of $X$, consider any outcome of the $n$ flips that has a total of $i$ heads and $n - i$ tails. For instance, consider the outcome $h, h, \ldots, h, t, t, \ldots, t$, in which the first $i$ flips result in heads and the others in tails. Since each flip is independent and results in heads with probability $p$, this outcome will arise with probability

$$p \cdot p \cdots p \cdot (1 - p) \cdots (1 - p) = p^i (1 - p)^{n-i}.$$

Indeed, it is easy to see that each outcome that has a total of $i$ heads and $n - i$ tails will also occur with probability $p^i (1 - p)^{n-i}$. Since there are $\binom{n}{i}$ outcomes that have a total of $i$ heads (most easily seen by noting that each such outcome is determined by the choice of which set of $i$ of the $n$ flips are to land on heads), it follows that

$$P\{X = i\} = \binom{n}{i} p^i (1 - p)^{n-i}, \quad i = 0, 1, \ldots, n.$$

As a check, note that

$$\sum_{i=0}^{n} P\{X = i\} = \sum_{i=0}^{n} \binom{n}{i} p^i (1 - p)^{n-i}$$

$$= (p + 1 - p)^n \quad \text{(by the binomial theorem)}$$

$$= 1. \qquad \square$$

Consider again an experiment with sample space $S = \{1, 2, \ldots, m\}$ and probabilities $p_i$ $(i = 1, \ldots, m)$. For a random variable $X$, we let $X(i)$ represent the value of $X$ when $i$ is the outcome of the experiment.

**Definition**    The *expected value* of $X$, denoted $E[X]$, is defined by

$$E[X] = \sum_{i \in S} X(i)p_i = \sum_{i=1}^{m} X(i)p_i.$$

Alternative names for $E[X]$ are the *expectation* or the *mean* of $X$.

**Example 3.5c**    If $X$ is the number of heads obtained when two fair coins are flipped, then

$$X(h, h) = 2, \quad X(h, t) = X(t, h) = 1, \quad X(t, t) = 0;$$

hence

$$E[X] = 2(1/4) + 1(1/4) + 1(1/4) + 0(1/4) = 1. \qquad \square$$

If $a$ and $b$ are constants then, since the random variable $aX + b$ assumes the value $aX(i) + b$ when $i$ is the outcome of the experiment, it follows that

$$E[aX + b] = \sum_{i \in S} [aX(i) + b]p_i$$

$$= a \sum_{i \in S} X(i)p_i + b \sum_{i \in S} p_i$$

$$= aE[X] + b. \qquad (3.12)$$

Suppose that the set of possible values of $X$ is $\{x_1, \ldots, , x_n\}$. We can also express $E[X]$ as a weighted average of these values. To do so, partition the sample space $S$ into the sets $S_j$ $(j = 1, \ldots, n)$, where $S_j$ is the set of all outcomes of the experiment that result in $X$ taking value $x_j$; that is,

$$S_j = \{i : X(i) = x_j\}.$$

We can now express $E[X]$ as follows:

$$E[X] = \sum_{i \in S} X(i) p_i$$

$$= \sum_{i \in \bigcup_j S_j} X(i) p_i \quad \text{(since } S = \bigcup_j S_j\text{)}$$

$$= \sum_{j=1}^{n} \sum_{i \in S_j} X(i) p_i \quad \text{(since the } S_j \text{ are disjoint)}$$

$$= \sum_{j=1}^{n} \sum_{i \in S_j} x_j p_i$$

$$= \sum_{j=1}^{n} x_j \sum_{i \in S_j} p_i$$

$$= \sum_{j=1}^{n} x_j P\{X = x_j\}.$$

Thus, we have proven the following.

**Proposition 3.5.1**

$$E[X] = \sum_{j=1}^{n} x_j P\{X = x_j\}.$$

In words, Proposition 3.5.1 states that $E[X]$ is a weighted average of the possible values of $X$, where the weight given to a value is equal to the probability that $X$ assumes that value.

**Example 3.5d**    Let the random variable $X$ denote the amount that we win when we make a certain bet. Find $E[X]$ if there is a 60% chance that we lose 1, a 20% chance that we win 1, and a 20% chance that we win 2.

*Solution.*

$$E[X] = -1(0.6) + 1(0.2) + 2(0.2) = 0.$$

Thus, the expected amount that is won on this bet is equal to 0. A bet whose expected winning is equal to 0 is called a *fair* bet.    $\square$

**Example 3.5e**  A random variable $X$ that is equal to 1 with probability $p$, and to 0 with probability $1 - p$, is said to be a *Bernoulli* random variable with parameter $p$. Its expected value is

$$E[X] = 1(p) + 0(1 - p) = p. \qquad \square$$

**Example 3.5f** *Utility*  Suppose that you must choose one of two possible actions, each of which can result in any of $n$ consequences, denoted as $C_1, \ldots, C_n$. Suppose that if the first action is chosen, then consequence $i$ will result with probability $p_i$ $(i = 1, \ldots, n)$, whereas if the second action is chosen, then consequence $i$ will result with probability $q_i$ $(i = 1, \ldots, n)$, where $\sum_{i=1}^{n} p_i = \sum_{i=1}^{n} q_i = 1$. The following approach can be used to determine which action to choose.

It starts by assigning numerical values to the different consequences in the following manner. First, identify the least and the most desirable consequence, call them $c$ and $C$ respectively; give the consequence $c$ the value 0 and give $C$ the value 1. Now consider any of the other $n - 2$ consequences, say $C_i$. To value this consequence, imagine that you are given the choice between either receiving $C_i$ or of taking part in a random experiment that either earns you consequence $C$ with probability $u$ or consequence $c$ with probability $1 - u$. Clearly your choice will depend on the value of $u$. If $u = 1$ then the experiment is certain to result in consequence $C$; since $C$ is the most desirable consequence, you will clearly prefer the experiment to receiving $C_i$. On the other hand, if $u = 0$ then the experiment will result in the least desirable consequence, namely $c$, so in this case you will clearly prefer the consequence $C_i$ to the experiment. Now, as $u$ decreases from 1 down to 0, it seems reasonable that your choice will at some point switch from the experiment to the certain return of $C_i$, and at that critical switch point you will be indifferent between the two alternatives. Take that indifference probability $u$ as the value of the consequence $C_i$. In other words, the value of $C_i$ is that probability $u$ such that you are indifferent between either receiving the consequence $C_i$ or taking part in an experiment that returns consequence $C$ with probability $u$ or consequence $c$ with probability $1 - u$. We call this indifference probability the *utility* of the consequence $C_i$, and we designate it as $u(C_i)$.

To determine which action is superior, we need to evaluate each one. Consider the first action, which results in consequence $C_i$ with probability $p_i$, $l = 1, \ldots, n$. We can think of the result of this action as being determined by a two-stage experiment. In the first stage, one of the values $1, \ldots, n$ is chosen according to the probabilities $p_1, \ldots, p_n$; if value $i$ is chosen, you receive consequence $C_i$. However, since $C_i$ is equivalent to obtaining consequence $C$ with probability $u(C_i)$ or consequence $c$ with probability $1 - u(C_i)$, it follows that the result of the two-stage experiment is equivalent to an experiment in which either consequence $C$ or $c$ is obtained, with $C$ being obtained with probability

$$\sum_{i=1}^{n} p_i u(C_i).$$

Similarly, the result of choosing the second action is equivalent to taking part in an experiment in which either consequence $C$ or $c$ is obtained, with $C$ being obtained with probability

$$\sum_{i=1}^{n} q_i u(C_i).$$

Since $C$ is preferable to $c$, it follows that the first action is preferable to the second action if

$$\sum_{i=1}^{n} p_i u(C_i) > \sum_{i=1}^{n} q_i u(C_i).$$

In other words, the value of an action can be measured by the expected value of the utility of its consequence, and the action with the largest expected utility is most preferable.  □

An important result is that the expected value of a sum of random variables is equal to the sum of their expected values.

**Proposition 3.5.2**  *For random variables $X_1, \ldots, X_k$,*

$$E\left[ \sum_{j=1}^{k} X_j \right] = \sum_{j=1}^{k} E[X_j].$$

***Proof.*** Because the random variable $\sum_{j=1}^{k} X_j$ will take on the value $\sum_{j=1}^{k} X_j(i)$ when $i$ is the outcome of the experiment, it follows that

$$E\left[\sum_{j=1}^{k} X_j\right] = \sum_{i \in S}\left(\sum_{j=1}^{k} X_j(i)\right) p_i$$

$$= \sum_{j=1}^{k} \sum_{i \in S} X_j(i) p_i$$

$$= \sum_{j=1}^{k} E[X_j]. \qquad \square$$

**Example 3.5g** Suppose that $n$ trials are performed, and that the $j$th trial is a success with probability $p_j$ ($j = 1, \ldots, n$). If $X$ represents the number of successes in these trials, then we can easily determine $E[X]$ by using the representation

$$X = \sum_{j=1}^{n} X_j,$$

where $X_j$ is defined to equal 1 if trial $j$ is a sucess and 0 otherwise. This equality yields

$$E[X] = \sum_{j=1}^{n} E[X_j]$$

$$= \sum_{j=1}^{n} p_j,$$

where the final equality used the result of Example 3.5d. $\qquad \square$

**Example 3.5h** Suppose that $n$ balls are randomly selected, without replacement, from an urn containing $N$ balls of which $m$ are white. Find the expected number of white balls chosen.

*Solution.* Let $X$ denote the number of white balls selected. Also, let

$$X_j = \begin{cases} 1 & \text{if the } j\text{th ball selected is white,} \\ 0 & \text{otherwise,} \end{cases}$$

and note that

$$X = \sum_{j=1}^{n} X_j.$$

Since each of the $N$ balls is equally likely to be the $j$th one selected, it follows that $X_j$ is a Bernoulli random variable with parameter

$$p = P\{X_j = 1\} = \frac{m}{N}.$$

Hence, from Example 3.5e,

$$E[X_j] = \frac{m}{N}$$

and thus

$$E[X] = \frac{nm}{N}. \qquad \square$$

The random variables $X_1, \ldots, X_n$ are said to be *independent* if probabilities concerning a subset of them are unchanged by information as to the values of the others.

**Example 3.5i**  Suppose that $k$ balls are to be randomly chosen from a set of $N$ balls of which $n$ are red. Let $X_i$ equal 1 if the $i$th ball chosen is red and 0 if it is black. Then $X_1, \ldots, X_n$ would be independent if each selected ball is replaced before the next selection is made, and they would not be independent if each selection is made without replacing previously selected balls. (Why not?) $\qquad \square$

Whereas the average of the possible values of $X$ is indicated by its expected value, its "spread" is measured by its variance.

**Definition**  The *variance* of $X$, denoted by $\mathrm{Var}(X)$, is defined by

$$\mathrm{Var}(X) = E[(X - E[X])^2].$$

In other words, the variance measures the average square of the difference between $X$ and its expected value.

**Example 3.5j**  Find $\mathrm{Var}(X)$ when $X$ is a Bernoulli random variable with parameter $p$.

*Solution.*  As shown in Example 3.5e, $E[X] = p$. Hence we see that

$$(X - E[X])^2 = \begin{cases} (1 - p)^2 & \text{with probability } p, \\ p^2 & \text{with probability } 1 - p. \end{cases}$$

It follows that

$$\text{Var}(X) = E[(X - E[X])^2]$$
$$= (1 - p)^2 p + p^2(1 - p)$$
$$= p - p^2. \qquad \square$$

If $a$ and $b$ are constants, then

$$\text{Var}(aX + b) = E[(aX + b - E[aX + b])^2]$$
$$= E[(aX - aE[X])^2] \quad \text{(by equation (3.12))}$$
$$= E[a^2(X - E[X])^2]$$
$$= a^2 \text{Var}(X). \qquad (3.13)$$

Although it is not generally true that the variance of the sum of random variables is equal to the sum of their variances, this *is* true when the random variables are independent.

**Proposition 3.5.3**  *If $X_1, \ldots, X_k$ are independent random variables, then*

$$\text{Var}\left(\sum_{j=1}^{k} X_j\right) = \sum_{j=1}^{k} \text{Var}(X_j).$$

**Example 3.5k**  Find the variance of $X$, a binomial random variable with parameters $n$ and $p$.

**Solution.**  Recalling that $X$ represents the number of successes in $n$ *independent* trials, each of which is a success with probability $p$, we can represent it as

$$X = \sum_{j=1}^{n} X_j,$$

where $X_j$ is defined to equal 1 if trial $j$ is a sucess and 0 otherwise. Hence,

$$\text{Var}(X) = \sum_{j=1}^{n} \text{Var}(X_j)$$

$$= \sum_{j=1}^{n} p(1 - p) \quad \text{(by Example 3.5j)}$$

$$= np(1 - p). \qquad \square$$

The square root of the variance is called the *standard deviation*.

## 3.6    Exercises

**Exercise 3.1**   When typing a report, a certain typist makes $i$ errors with probability $p_i$ $(i \geq 0)$, where

$$p_0 = 0.20, \qquad p_1 = 0.35,$$
$$p_2 = 0.25, \qquad p_3 = 0.15.$$

What is the probability that the typist makes

(a) at least 4 errors;
(b) at most 2 errors?

**Exercise 3.2**   A family picnic scheduled for tomorrow will be postponed if it is either cloudy or rainy. If the probability that it will be cloudy is 0.40, the probability that it will be rainy is 0.30, and the probability that it will be both rainy and cloudy is 0.20, then what is the probabilty that the picnic will not be postponed?

**Exercise 3.3**   If two people are randomly chosen from a group of eight women and six men, what is the probability that

(a) both are women;
(b) both are men;
(c) one is a man and the other a woman?

**Exercise 3.4**   A club has 120 members, of whom 35 play chess, 58 play bridge, and 27 play both chess and bridge. If a member of the club is randomly chosen, what is the conditional probability that she

(a) plays chess given that she plays bridge;
(b) plays bridge given that she plays chess?

**Exercise 3.5**   Cystic fibrosis (CF) is a genetically caused disease. A child that receives a CF gene from each of its parents will develop the disease either as a teen-ager or before and will not live to adulthood. A child that receives either zero or one CF gene will not develop the disease. If an individual has a CF gene, then each of his or her children will independently receive that gene with probability $1/2$.

(a) If both parents possess the CF gene, what is the probability that their child will develop cystic fibrosis?
(b) What is the probability that a 30-year old who does not have cystic fibrosis, but whose sibling died of that disease, possesses a CF gene?

**Exercise 3.6** Two cards are randomly selected from a deck of 52 playing cards. What is the conditional probability they are both aces, given that they are of different suits?

**Exercise 3.7** If $A$ and $B$ are independent, show that so are

(a) $A$ and $B^c$;
(b) $A^c$ and $B^c$.

**Exercise 3.8** You have asked a friend to water a sick plant while you are on vacation. Without water, it will die with probability 0.7; with water, it will die with probability 0.1. You are 90% certain that your friend will remember to water the plant.

(a) What is the probability that the plant will be alive when you return?
(b) If it is dead, what is the probability that your friend forgot to water it?

**Exercise 3.9** A cancer diagnostic test is 95% accurate both on those that do and those that do not have the cancer. If 0.4% of the population unknowingly has the cancer, what is the probability that a randomly chosen person who tests positive actually has the cancer?

**Exercise 3.10** An insurance company estimates that the probability that a certain event will occur in the coming year is $p$. A customer wants to purchase an insurance policy that will return the amount $R$ if the event occurs. How much should the insurance company charge for such a policy in order that its expected profit be 20% of $R$?

**Exercise 3.11** A gambling book recommends the following "winning strategy" for the game of roulette. It recommends that the gambler bet 1 on red. If red appears (which has probability 18/38 of occurring) then the gambler should take his profit of 1 and quit. If the gambler loses this bet, then he should then make a second bet of size 2 and then quit. Let $X$ denote the gambler's winnings.

(a) Find $P\{X > 0\}$.
(b) Find $E[X]$.

**Exercise 3.12**   Four buses carrying 152 students from the same school arrive at a football stadium. The buses carry (respectively) 39, 33, 46, and 34 students. One of the 152 students is randomly chosen; let $X$ denote the number of students who were on the bus of the selected student. One of the four bus drivers is also randomly chosen; let $Y$ be the number of students who were on that driver's bus.

(a) Which do you think is larger, $E[X]$ or $E[Y]$?
(b) Find $E[X]$ and $E[Y]$.

**Exercise 3.13**   There are $k$ different types of coupons, and each one obtained will independently be of type $i$ with probability $p_i$, $i = 1, \ldots, k$. Find the expected number of different types that are contained in a set of $n$ coupons.

**Exercise 3.14**   A group of $n$ people throw their hats into the center of a room. The hats are scrambled and each person chooses one. If each then puts the selected hat on his or her head, what is the expected number that are wearing their own hat?

**Exercise 3.15**   Two players play a tennis match, which ends when one of the players has won two sets. Suppose that each set is equally likely to be won by either player, and that the results from different sets are independent. Find (a) the expected value and (b) the variance of the number of sets played.

**Exercise 3.16**   A lawyer must decide whether to charge a fixed fee of $5,000 or take a contingency fee of $25,000 if she wins the case (and 0 if she loses). She estimates that her probability of winning is 0.30. Determine the mean and standard deviation of her fee if

(a) she takes the fixed fee;
(b) she takes the contingency fee.

# 4. Mathematics of Finance

## 4.1 Interest Rates

If you borrow the amount $P$ (called the principal) which must be repaid at time $T$ along with simple interest at rate $r$ per time $T$, then the amount to be repaid at time $T$ is

$$P + rP = (1 + r)P.$$

That is, you must repay both the principal $P$ and the interest, equal to the principal times the interest rate. For instance, if you borrow $100 which is to be repaid after one year with a simple interest rate of 5% per year (i.e., $r = 0.05$), then you will have to repay $105 at the end of the year.

**Example 4.1a** Suppose that you borrow an amount $P$ that is to be repaid after one year along with interest at a rate $r$ per year *compounded* semiannually. What does this mean? How much is owed in a year?

*Solution.* In order to answer this, it is necessary to know that having your interest compounded semiannually means that after half a year you will be charged simple interest at the rate of $r/2$ per half-year; that interest is then added on to your principal, which is again charged interest at rate $r/2$ for the second half-year period. In other words, after six months you owe

$$P(1 + r/2).$$

This is then regarded as the new principal for another six-month loan at interest rate $r/2$. Hence, at the end of the year you will owe

$$P(1 + r/2)(1 + r/2) = P(1 + r/2)^2. \qquad \square$$

**Example 4.1b** If you borrow $1,000 for one year at an interest rate of 8% per year that is to be compounded quarterly, how much do you owe at the end of the year?

**Solution.** An interest rate of 8% that is compounded quarterly is equivalent to paying simple interest at a rate of 2% per quarter year, with each additional quarter charging interest not only on the original principal but also on the interest that has accrued up to that point. Thus, after one quarter you owe

$$1{,}000(1 + 0.02);$$

after two quarters you owe

$$1{,}000(1 + 0.02)(1 + 0.02) = 1{,}000(1 + 0.02)^2;$$

after three quarters you owe

$$1{,}000(1 + 0.02)^2(1 + 0.02) = 1{,}000(1 + 0.02)^3;$$

and after four quarters you owe

$$1{,}000(1 + 0.02)^3(1 + 0.02) = 1{,}000(1 + 0.02)^4 = 1{,}082.40. \quad \square$$

**Example 4.1c**   Many credit-card companies charge interest at a yearly rate of 18% compounded monthly. If the amount $P$ is charged at the beginning of a year, how much is owed at the end of the year if no previous payments have been made?

**Solution.** Such a compounding is equivalent to paying simple interest every month at a rate of $18/12 = 1.5\%$ per month, with the accrued interest then added to the principal owed during the next month. Hence, after one year you will owe

$$P(1 + 0.015)^{12} = 1.1956P. \quad \square$$

If the interest rate $r$ is compounded then, as we have seen in Examples 4.1b and 4.1c, the amount of interest actually paid is greater than if we were paying simple interest at rate $r$. The reason, of course, is that in compounding we are being charged interest on the interest that has already been computed in previous compoundings. In these cases, we call $r$ the *nominal* interest rate, and we define the *effective interest rate*, call it $r_{\text{eff}}$, by

$$r_{\text{eff}} = \frac{\text{amount repaid at the end of a year} - P}{P}.$$

For instance, if the loan is for one year at a nominal interest rate $r$ that is to be compounded quarterly, then the effective interest rate for the year is

$$r_{\text{eff}} = (1 + r/4)^4 - 1.$$

Thus, in Example 4.1b the effective interest rate is 8.24% whereas in Example 4.1c it is 19.56%. Since

$$P(1 + r_{\text{eff}}) = \text{amount repaid at the end of a year,}$$

the payment made in a one-year loan with compound interest is the same as if the loan called for simple interest at rate $r_{\text{eff}}$ per year.

Suppose now that we borrow the principal $P$ for one year at a nominal interest rate of $r$ per year, compounded *continuously*. Now, how much is owed at the end of the year? Of course, before answering this we must decide on an appropriate definition of "continuous" compounding. Note that, if the loan is compounded at $n$ equal intervals in the year, then the amount owed at the end of the year is $P(1 + r/n)^n$. Thus, since it is reasonable to suppose that continuous compounding refers to the limit of the preceding as $n$ grows larger and larger, the amount owed at time 1 is

$$P \lim_{n \to \infty} (1 + r/n)^n = Pe^r,$$

where $e$, the base of the natural logarithm, is defined by

$$e = \lim_{n \to \infty} (1 + 1/n)^n$$

and is approximately given by $e \approx 2.71828 \ldots$.

**Example 4.1d**    If a bank offers interest at a nominal rate of 5% compounded continuously, what is the effective interest rate per year?

*Solution.*  The effective interest rate is

$$r_{\text{eff}} = \frac{Pe^{0.05} - P}{P} = e^{0.05} - 1 \approx 0.05127.$$

That is, the effective interest rate is 5.127% per year.    □

If the amount $P$ is borrowed for $t$ years at a nominal interest rate of $r$ per year compounded continuously, then the amount owed at time $t$ is

$Pe^{rt}$. This is seen by interpreting the interest rate as being a continuous compounding of a nominal rate of $tr$ per time $t$; hence, the amount owed at time $t$ is

$$P \lim_{n \to \infty} (1 + rt/n)^n = Pe^{rt}.$$

## 4.2    Present Value Analysis

Suppose that one can both borrow and loan money at a nominal rate $r$ that is compounded periodically. Under these conditions, what is the present worth of a payment of $v$ dollars that will be made at the end of period $i$? Since a bank loan of $v(1+r)^{-i}$ would require a payoff of $v$ at period $i$, it follows that the *present value* of a payoff of $v$ to be made at time period $i$ is $v(1+r)^{-i}$.

The concept of present value enables us to compare different income streams to see which is preferable.

**Example 4.2a**    Suppose that you are to receive payments (in thousands of dollars) at the end of each of the next five years. Which of the following three payment sequences is preferable?

**A.** 12, 14, 16, 18, 20;
**B.** 16, 16, 15, 15, 15;
**C.** 20, 16, 14, 12, 10.

**Solution.** If the present nominal interest rate is $r$ compounded yearly, then the present value of the sequence of payments $x_i$ $(i = 1, 2, 3, 4, 5)$ is

$$\sum_{i=1}^{5} (1 + r)^{-i} x_i;$$

the sequence having the largest present value is preferred. It thus follows that the superior sequence of payments depends on the interest rate. If $r$ is small, then the sequence **A** is best since its sum of payments is the highest. For a somewhat larger value of $r$, the sequence **B** would be best because – although the total of its payments (77) is less than that of **A** (80) – its earlier payments are larger than are those of **A**. For an even larger value of $r$, the sequence **C**, whose earlier payments are higher than those in either **A** or **B**, would be best. Table 4.1 gives the present values of these payment streams for three different values of $r$.

Table 4.1: *Present Values*

|  | Payment Sequence | | |
| --- | --- | --- | --- |
| $r$ | A | B | C |
| 0.1 | 59.21 | 58.60 | 56.33 |
| 0.2 | 45.70 | 46.39 | 45.69 |
| 0.3 | 36.49 | 37.89 | 38.12 |

It should be noted that the payment sequences can be compared according to their values at any specified time. For instance, if we compare them according to their time-5 values, then we would determine which sequence of payments yields the largest value of

$$\sum_{i=1}^{5}(1+r)^{5-i}x_i = (1+r)^5 \sum_{i=1}^{5}(1+r)^{-i}x_i;$$

this is the same sequence choice as before. □

**Example 4.2b** Suppose that one takes a mortgage loan for the amount $L$ that is to be paid back over $n$ months with equal payments of $A$ at the end of each month. The interest rate for the loan is $r$ per month, compounded monthly.

(a) In terms of $L$, $n$, and $r$, what is the value of $A$?
(b) After payment has been made at the end of month $j$, how much additional loan principal remains?
(c) How much of the payment during month $j$ is for interest and how much is for principal reduction? (This is important because some contracts allow for the loan to be paid back early and because the interest part of the payment is tax-deductible.)

*Solution.* The present value of the $n$ monthly payments is

$$\frac{A}{1+r} + \frac{A}{(1+r)^2} + \cdots + \frac{A}{(1+r)^n}$$

$$= \frac{A}{1+r}\left[1 + \frac{1}{1+r} + \left(\frac{1}{1+r}\right)^2 + \cdots + \left(\frac{1}{1+r}\right)^{n-1}\right]$$

$$= \frac{A}{1+r} \frac{1 - \left(\frac{1}{1+r}\right)^n}{1 - \frac{1}{1+r}}$$

$$= \frac{A}{r}[1 - (1+r)^{-n}].$$

Since this must equal the loan amount $L$, we see that

$$A = \frac{Lr}{1 - (1+r)^{-n}} = \frac{L(\alpha - 1)\alpha^n}{\alpha^n - 1}, \qquad (4.1)$$

where

$$\alpha = 1 + r.$$

For instance, if the loan is for \$100,000, to be paid back over 360 months at a nominal yearly interest rate of 0.09 compounded monthly, then $r = 0.09/12 = 0.0075$ and the monthly payment (in dollars) would be

$$A = \frac{100,000(0.0075)(1.0075)^{360}}{(1.0075)^{360} - 1} = 804.62.$$

Let $R_j$ denote the remaining amount of principal owed after the payment at the end of month $j$, $j = 0, \ldots, n$. To determine these quantities, note that if one owes $R_j$ at the end of month $j$ then the amount owed immediately before the payment at the end of month $j + 1$ is $(1+r)R_j$; one then pays the amount $A$, so it follows that

$$R_{j+1} = (1+r)R_j - A = \alpha R_j - A.$$

Starting with $R_0 = L$, we obtain:

$$R_1 = \alpha L - A,$$

$$R_2 = \alpha R_1 - A$$

$$= \alpha(\alpha L - A) - A$$

$$= \alpha^2 L - (1 + \alpha)A,$$

$$R_3 = \alpha R_2 - A$$

$$= \alpha(\alpha^2 L - (1 + \alpha)A) - A$$

$$= \alpha^3 L - (1 + \alpha + \alpha^2)A.$$

In general, we obtain

$$R_j = \alpha^j L - A(1 + \alpha + \cdots + \alpha^{j-1})$$

$$= \alpha^j L - A\frac{\alpha^j - 1}{\alpha - 1}$$

$$= \alpha^j L - \frac{L\alpha^n(\alpha^j - 1)}{\alpha^n - 1} \quad \text{(from (4.1))}$$

$$= \frac{L(\alpha^n - \alpha^j)}{\alpha^n - 1}, \quad j = 0, \ldots, n.$$

Let $I_j$ and $P_j$ denote (respectively) the amounts of the payment at the end of month $j$ that are for interest and for principal reduction. Then, since $R_{j-1}$ was owed at the end of the previous month, we have

$$I_j = rR_{j-1}$$

$$= \frac{L(\alpha - 1)(\alpha^n - \alpha^{j-1})}{\alpha^n - 1}$$

and

$$P_j = A - I_j$$

$$= \frac{L(\alpha - 1)}{\alpha^n - 1}[\alpha^n - (\alpha^n - \alpha^{j-1})]$$

$$= \frac{L(\alpha - 1)\alpha^{j-1}}{\alpha^n - 1}.$$

As a check, note that

$$\sum_{j=1}^{n} P_j = L.$$

It follows from the preceding that the amount of principal repaid in subsequent months increases by the factor $\alpha = 1 + r$. For example, in a $100,000 loan for 30 years at a nominal interest rate of 9% per year compounded monthly, only $54.62 of the $804.62 paid during the first month goes toward reducing the principal of the loan; the remainder is interest. In each succeeding month, the amount of the payment that goes toward principal increases by the factor 1.0075. □

**Example 4.2c** An individual who plans to retire in 20 years has decided to put an amount $A$ in the bank at the beginning of each of the next

240 months, after which she will withdraw $1,000 at the beginning of each of the following 360 months. Assuming a nominal yearly interest rate of 6% compounded monthly, how large does $A$ need to be?

*Solution.* Let $r = 0.06/12 = 0.005$ be the monthly interest rate. With $\beta = \frac{1}{1+r}$, the present value of all her deposits is

$$A + A\beta + A\beta^2 + \cdots + A\beta^{239} = A\frac{1 - \beta^{240}}{1 - \beta}.$$

Similarly, if $W$ is the amount withdrawn in the following 360 months, then the present value of all these withdrawals is

$$W\beta^{240} + W\beta^{241} + \cdots + W\beta^{599} = W\beta^{240}\frac{1 - \beta^{360}}{1 - \beta}.$$

Thus she will be able to fund all withdrawals (and have no money left in her account) if

$$A\frac{1 - \beta^{240}}{1 - \beta} = W\beta^{240}\frac{1 - \beta^{360}}{1 - \beta}.$$

With $W = 1,000$ and $\beta = 1/1.005$, this gives

$$A = 360.99.$$

That is, saving $361 a month for 240 months will enable her to withdraw $1,000 a month for the succeeding 360 months. $\square$

## 4.3    Pricing Contracts via Arbitrage

### 4.3.1    *An Example in Options Pricing*

Suppose that the nominal interest rate is $r$, and consider the following model for purchasing a stock at a future time at a fixed price. Let the present price (in dollars) of the stock be 100 per share, and suppose we know that, after one time period, its price will be either 200 or 50 (see Figure 4.1). Suppose further that, for any $y$, at a cost of $cy$ you can purchase at time 0 the option to buy $y$ shares of the stock at time 1 at a price of 150 per share. Thus, for instance, if you purchase this option and the

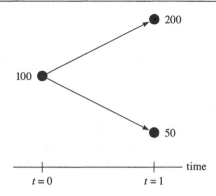

$t = 0$      $t = 1$

Figure 4.1:  Possible Stock Prices at Time 1

stock rises to 200, then you would exercise the option at time 1 and realize a gain of $200 - 150 = 50$ for each of the $y$ options purchased. On
the other hand, if the price of the stock at time 1 is 50, then the option
would be worthless. In addition to the options, you may also purchase $x$
shares of the stock at time 0 at a cost of of $100x$, and each share would
be worth either 200 or 50 at time 1.

We will suppose that both $x$ and $y$ can be positive, negative, or zero.
That is, you can either buy or sell both the stock and the option. For instance, if $x$ were negative then you would be selling $-x$ shares of stock,
yielding you an initial return of $-100x$; you would then be responsible
for buying and returning $-x$ shares of the stock at time 1 at a (time-1)
cost of either 200 or 50 per share. (When you sell a stock that you do
not own, we say that you are selling it *short*.)

We are interested in the appropriate value of $c$, the unit cost of an option. Specifically, we will show that, unless $c = [100 - 50(1 + r)^{-1}]/3$,
there is a combination of purchases that will always result in a positive
gain. To show this, suppose that at time 0 we

(a)  purchase $x$ units of stock, and
(b)  purchase $y$ units of options,

where $x$ and $y$ (both of which can be either positive or negative) are to
be determined. The cost of this transaction is $100x + cy$; if this amount
is positive, then it should be borrowed from a bank, to be repaid with interest at time 1; if it is negative, then the amount received, $-(100x + cy)$,

should be put in the bank to be withdrawn at time 1. The value of our holdings at time 1 depends on the price of the stock at that time and is given by

$$\text{value} = \begin{cases} 200x + 50y & \text{if the price is 200,} \\ 50x & \text{if the price is 50.} \end{cases}$$

This formula follows by noting that, if the stock's price at time 1 is 200, then the $x$ shares of the stock are worth $200x$ and the $y$ units of options to buy the stock at a share price of 150 are worth $(200 - 150)y$. On the other hand, if the stock's price is 50, then the $x$ shares are worth $50x$ and the $y$ units of options are worthless. Now, suppose we choose $y$ so that the value of our holdings at time 1 is the same no matter what the price of the stock at that time. That is, we choose $y$ so that

$$200x + 50y = 50x$$

or

$$y = -3x.$$

Note that $y$ has the opposite sign from $x$; thus, if $x > 0$ and so $x$ shares of the stock are purchased at time 0, then $3x$ units of stock options are also *sold* at that time. Similarly, if $x$ is negative, then $-x$ shares are sold and $-3x$ units of stock options are purchased at time 0.

Thus, with $y = -3x$, the value of our holdings at time 1 is

$$\text{time-1 value of holdings} = 50x$$

no matter what the value of the stock. As a result, if $y = -3x$ it follows that, after paying off our loan (if $100x + cy > 0$) or withdrawing our money from the bank (if $100x + cy < 0$), we will have gained the amount

$$\begin{aligned} \text{gain} &= 50x - (100x + cy)(1 + r) \\ &= 50x - (100x - 3xc)(1 + r) \\ &= (1 + r)x[3c - 100 + 50(1 + r)^{-1}]. \end{aligned}$$

Thus, if $3c = 100 - 50(1 + r)^{-1}$ then the gain is 0; on the other hand, if $3c \neq 100 - 50(1+r)^{-1}$ then we can guarantee a positive gain (no matter what the price of the stock at time 1) by letting $x$ be positive when $3c > 100 - 50(1+r)^{-1}$ and letting $x$ be negative when $3c < 100 - 50(1+r)^{-1}$.

For instance, if $(1 + r)^{-1} = 0.9$ and the cost per option is $c = 20$, then purchasing one share of the stock and selling three units of options initially costs us $100 - 3(20) = 40$, which is borrowed from the bank. However, the value of this holding at time 1 is 50 whether the stock price rises to 200 or falls to 50. Using $40(1 + r) = 44.44$ of this amount to pay our bank loan results in a guaranteed gain of 5.56. Similarly, if the cost of an option is 15, then selling one share of the stock ($x = -1$) and buying three units of options results in an initial gain of $100 - 45 = 55$, which is put into a bank to be worth $55(1 + r) = 61.11$ at time 1. The value of our holding at time 1 is $-50$, so a guaranteed profit of 11.11 is attained. A sure-win betting scheme is called an *arbitrage*. Thus, for the numbers considered, the only option cost $c$ that does not result in an arbitrage is $c = (100 - 45)/3 = 55/3$.

### 4.3.2 *Other Examples of Pricing via Arbitrage*

The type of option we have been considering is known as a *call* option because it gives one the option of calling for the stock at a specified price, known as the *exercise* price. An *American style* call option allows the buyer to exercise the option at any time up to the expiration time, whereas a *European style* call option can only be exercised at the expiration time. Although it might seem that, because of its additional flexibility, the American style option should be worth more, it turns out that it is never optimal to exercise a call option early; thus, the two styles of options have identical worths. We now prove this result.

**Proposition 4.3.1** *One should never exercise an American style call option before its expiration time $t$.*

*Proof.* Suppose that the present price of the stock is $S$; that you own an option to buy one share of the stock at a fixed price $K$; and that the option expires after an additional time $t$. If you exercise the option then you will realize the amount $S - K$. However, consider what would transpire if, instead of exercising the option, you sell the stock short and then purchase the stock at time $t$, either by paying the market price at that time or by exercising your option and paying $K$, whichever is less expensive. Under this strategy, you will initially receive $S$ and will then have to pay the minimum of the market price and the exercise price $K$

after an additional time $t$. Clearly this dominates receiving $S$ and immediately paying out $K$.                                                    □

In addition to call options there are also *put* options on stocks. These
give their owners the option of putting a stock up for sale at a specified
price. An American style put option allows the owner to put the stock up
for sale – that is, to exercise the option – at any time up to the expiration
time of the option. A European style put option can only be exercised
at its expiration time. Contrary to the situation with call options, it may
be advantageous to exercise a put option before its expiration time and
so the American style put option may be worth more than the European.
The absence of arbitrage implies a relationship between the price of a
European put option having exercise price $K$ and expiration time $t$ and
the price of a call option on that stock that also has exercise price $K$ and
expiration time $t$. This is known as the *put–call option parity formula*.

**Proposition 4.3.2**    *Let C be the price of a call option that enables its
holder to buy one share of a stock at an exercise price K at time t; also,
let P be the price of a European put option that enables its holder to sell
one share of the stock for the amount K at time t. Let S be the price of the
stock at time 0. Then, assuming that interest is continuously discounted
at a nominal rate r, either*

$$S + P - C = Ke^{-rt}$$

*or there is an arbitrage opportunity.*

**Proof.**  If

$$S + P - C < Ke^{-rt},$$

then we can effect a sure win by initially buying one share of the stock,
buying one put option, and selling one call option. This initial payout
of $S + P - C$ is borrowed from a bank to be repaid at time $t$. Let us
now consider the value of our holdings at time $t$. There are two cases,
depending on $S(t)$, the stock's market price at time $t$. If $S(t) \leq K$ then
the call option we sold is worthless and we can exercise our put option
to sell the stock for the amount $K$. On the other hand, if $S(t) > K$ then
our put option is worthless and the call option we sold will be exercised,
forcing us to sell our stock for the price $K$. Thus, in either case we will

realize the amount $K$ at time $t$. Since $K > e^{rt}(S + P - C)$, we can pay off our bank loan and realize a positive profit in all cases.

When

$$S + P - C > Ke^{-rT},$$

we can make a sure profit by reversing what we did in the previous case. Namely, we now sell one share of stock, sell one put option, and buy one call option. We leave the details of the verification in this case as an exercise. □

The arbitrage principle also determines the relationship between the present price of a stock and the contracted price to buy the stock at a specified time in the future. Our next two examples are related to these so-called forwards contracts.

**Example 4.3a** *Forwards Contracts* Let $S$ be the present market price of a specified stock. In a forwards agreement, one agrees at time 0 to pay $F$ at time $t$ for one share of the stock that will be delivered at time $t$. That is, one contracts a price for the stock, which is to be delivered and paid for at time $t$. We will now give an arbitrage argument to show that, if interest is continuously discounted at the nominal interest rate $r$, then in order for there to be no arbitrage opportunity we must have

$$F = Se^{rt}.$$

To see why the preceding must hold, suppose that instead

$$F < Se^{rt}.$$

Then a sure win is obtained by selling the stock at time 0, with the understanding that you will buy it back at time $t$. Put the sale proceeds $S$ into a bond that matures at time $t$ and, in addition, buy a forwards contract for delivery of one share of the stock at time $t$. Thus, at time $t$ you will receive $Se^{rt}$ from your bond. From this, you pay $F$ to obtain one share of the stock, which you then return to settle your obligation. You thus end with a positive profit of $Se^{rt} - F$. On the other hand, if

$$F > Se^{rt}$$

then you can guarantee a profit of $F - Se^{rt}$ by simultaneously selling a forwards contract and borrowing $S$ to purchase the stock. At time $t$ you will receive $F$ for your stock, out of which you repay your loan amount of $Se^{rt}$.                                                    □

When one purchases a share of a stock in the stock market, one is purchasing a share of ownership in the entity that issues the stock. On the other hand, the commodity market deals with more concrete objects: agricultural items like oats, corn, and wheat; energy products like crude oil and natural gas; metals such as gold, silver, and platinum; animal parts such as hogs, pork-bellies, and beef; and so on. Almost all activity on the commodities market is involved with contracts for future purchases and sales of the commodity. Thus, for instance, you could purchase a contract to buy natural gas in 90 days for a price that is specified today. (Such a *futures contract* differs from a forwards contract in that, although both require payment in full when delivery is taken, in futures contracts one also settles up on a daily basis depending on the change of the price of the futures contract on the commodity exchange.) You could also write a futures contract that obligates you to sell gas at a specified price at a specified time. Most people that play the commodities market never have actual contact with the commodity. Rather, people that buy a futures contract most often sell that contract before the delivery date.

The relationship given in Example 4.3a does not hold for futures contracts in the commodity market. For one thing, if $F > Se^{rt}$ and you purchase the commodity (say, crude oil) to sell back at time $t$, then you will incur additional costs related to storing and insuring the oil. Also, when $F < Se^{rt}$, to sell the commodity for today's price you must be able to deliver it immediately.

One of the most popular types of future contracts involves currency exchanges. The following example deals with this topic.

**Example 4.3b**  The *New York Times* gives the following listing for the price of a German mark (or DM):

- today – 0.5906;
- 90-day forward – 0.5921.

In other words, you can purchase 1 DM today at the price of $0.5906. You can also sign a contract to purchase 1 DM in 90 days at a price, to be paid on delivery, of $0.05921. Why are these prices different?

**Solution.** One might suppose that the difference is caused by the market's expectation of the worth in 90 days of the German DM relative to the U.S. dollar. However, it turns out that the entire price differential is due to the different interest rates in Germany and in the United States. Suppose that interest in both countries is continuously compounded at nominal yearly rates: $r_u$ in the United States and $r_g$ in Germany. Let $S$ denote the present price of one DM, and let $F$ be the price for a futures contract to be delivered at time $t$. (The example considers the special case where $S = 0.5906$, $F = 0.5921$, and $t = 90/365$.) We now argue that, in order for there not to be an arbitrage opportunity, we must have

$$F = Se^{(r_u - r_g)t}.$$

To see why, suppose first that

$$Fe^{r_g t} > Se^{r_u t}.$$

To obtain a sure win, borrow $S$ dollars to be repaid at time $t$. Use these dollars to buy 1 DM, which is used in turn to buy a German bond that matures at time $t$. (Thus, at time $t$ you will have $e^{r_g t}$ German marks and you will owe an American bank $Se^{r_u t}$ dollars.) Also, write a contract to sell $e^{r_g t}$ German marks at time $t$ for a total price of $Fe^{r_g t}$ dollars. Then, at time $t$ you sell your German marks for $Fe^{r_g t}$ dollars, use $Se^{r_u t}$ of this to pay off your American debt, and end with a profit of $Fe^{r_g t} - Se^{r_u t}$.

Suppose now that

$$Fe^{r_g t} < Se^{r_u t};$$

then you can obtain a sure win by reversing the preceding operation as follows. At time 0, buy a futures contract to purchase $e^{r_g t}$ DM at time $t$; borrow 1 DM from a German bank and sell it for $S$ dollars, which you then use to buy an American bond maturing at time $t$. Thus, at time $t$ you will have $Se^{r_u t}$ dollars; use $Fe^{r_g t}$ of it to pay for your futures contract. This gives you $e^{r_g t}$ marks, which you then use to retire your German bank loan debt. Hence, you end with a profit of $Se^{r_u t} - Fe^{r_g t}$.    □

## 4.4    The Arbitrage Theorem

Consider an experiment whose set of possible outcomes is $\{1, 2, \ldots, m\}$, and suppose that $n$ wagers concerning this experiment are available. If the amount $x$ is bet on wager $i$, then the amount $xr_i(j)$ is received if the outcome of the experiment is $j$ ($j = 1, \ldots, m$). In other words, $r_i(\cdot)$ is the "return function" for a unit bet on wager $i$. The amount bet on a wager is allowed to be positive, negative, or zero.

A betting strategy is a vector $\mathbf{x} = (x_1, x_2, \ldots, x_n)$, with the interpretation that $x_1$ is bet on wager 1, $x_2$ is bet on wager 2, $\ldots$, $x_n$ is bet on wager $n$. If the outcome of the experiment is $j$, then the return from the betting strategy $\mathbf{x}$ is given by

$$\text{return from } \mathbf{x} = \sum_{i=1}^{n} x_i r_i(j).$$

The following result, which is known as the *arbitrage theorem,* states that either there exists a probability vector $\mathbf{p} = (p_1, p_2, \ldots, p_m)$ on the set of possible outcomes of the experiment under which the expected return of each wager is equal to 0, or else there exists a betting strategy that yields a positive win for each outcome of the experiment.

**Theorem 4.4.1** (Arbitrage Theorem)  *Exactly one of the following is true: Either*

(a) *there is a probability vector $\mathbf{p} = (p_1, p_2, \ldots, p_m)$ for which*

$$\sum_{j=1}^{m} p_j r_i(j) = 0 \quad \text{for all } i = 1, \ldots, n,$$

*or else*

(b) *there is a betting strategy $\mathbf{x} = (x_1, x_2, \ldots, x_n)$ for which*

$$\sum_{i=1}^{n} x_i r_i(j) > 0 \quad \text{for all } j = 1, \ldots, m.$$

If $X$ is the outcome of the experiment, then the arbitrage theorem states that either there is a set of probabilities $(p_1, p_2, \ldots, p_m)$ such that if

$$P\{X = j\} = p_j \quad \text{for all } j = 1, \ldots, m$$

then

$$E[r_i(X)] = 0 \quad \text{for all } i = 1, \ldots, n,$$

or else there is a betting strategy that leads to a sure win. In other words, either there is a probability vector on the outcomes of the experiment that results in all bets being fair, or else there is a betting scheme that guarantees a win.

A proof of the arbitrage theorem, using the duality theorem of linear programming, will be presented in Section 7.3.

**Example 4.4a**   In some situations, the only type of wagers allowed are ones that choose one of the outcomes $i$ $(i = 1, \ldots, m)$ and then bet that $i$ is the outcome of the experiment. The return from such a bet is often quoted in terms of *odds*. If the odds against outcome $i$ are $o_i$ (often expressed as "$o_i$ to 1"), then a one-unit bet will return either $o_i$ if $i$ is the outcome of the experiment or $-1$ if $i$ is not the outcome. That is, a one-unit bet on $i$ will either win $o_i$ or lose 1. The return function for such a bet is given by

$$r_i(j) = \begin{cases} o_i & \text{if } j = i, \\ -1 & \text{if } j \neq i. \end{cases}$$

Suppose that the odds $o_1, o_2, \ldots, o_m$ are quoted. In order for there not to be a sure win, there must be a probability vector $\mathbf{p} = (p_1, p_2, \ldots, p_m)$ such that, for each $i$ $(i = 1, \ldots, m)$,

$$0 = E_{\mathbf{p}}[r_i(X)] = o_i p_i - (1 - p_i).$$

That is, we must have

$$p_i = \frac{1}{1 + o_i}.$$

Since the $p_i$ must sum to 1, this means that the condition for there not to be an arbitrage is that

$$\sum_{i=1}^{m} \frac{1}{1 + o_i} = 1.$$

This means that if $\sum_{i=1}^{m}(1 + o_i)^{-1} \neq 1$ then a sure win is possible.

For instance, suppose there are three possible outcomes and the quoted odds are as follows.

| Outcome | Odds |
| --- | --- |
| 1 | 1 |
| 2 | 2 |
| 3 | 3 |

That is, the odds against outcome 1 are 1 to 1; they are 2 to 1 against outcome 2 and they are 3 to 1 against outcome 3. Since

$$\frac{1}{2} + \frac{1}{3} + \frac{1}{4} = \frac{13}{12} \neq 1,$$

a sure win is possible. One possibility is to bet $-1$ on outcome 1 (so you either win 1 if the outcome is not 1 or you lose 1 if the outcome is 1) and also bet $-0.7$ on outcome 2 (so you either win 0.7 if the outcome is not 2 or you lose 1.4 if it is 2) and $-0.5$ on outcome 3 (so you either win 0.5 if the outcome is not 3 or you lose 1.5 if it is 3). If the experiment results in outcome 1 then you win $-1 + 0.7 + 0.5 = 0.2$; if it results in outcome 2, you win $1 - 1.4 + 0.5 = 0.1$; if it results in outcome 3, you win $1 + 0.7 - 1.5 = 0.2$. Hence, in all cases you win a positive amount. □

**Example 4.4b**  Let us reconsider the option pricing example of Section 4.3.1, where the initial price of a stock is 100 and the price after one period is assumed to be either 200 or 50. At a cost of $c$ per share, we can purchase at time 0 the option to buy the stock at time 1 for the price of 150. For what value of $c$ is no sure win possible?

*Solution.*  In the context of this section, the outcome of the experiment is the value of the stock at time 1; thus, there are two possible outcomes. There are also two different wagers: to buy (or sell) the stock, and to buy (or sell) the option. By the arbitrage theorem, there will be no sure win if there are probabilities $(p, 1 - p)$ on the outcomes that make the expected present value return equal to zero for both wagers.

The present value return from purchasing one share of the stock is

$$\text{return} = \begin{cases} 200(1 + r)^{-1} - 100 & \text{if the price is 200 at time 1,} \\ 50(1 + r)^{-1} - 100 & \text{if the price is 50 at time 1.} \end{cases}$$

Hence, if $p$ is the probability that the price is 200 at time 1, then

$$E[\text{return}] = p\left[\frac{200}{1 + r} - 100\right] + (1 - p)\left[\frac{50}{1 + r} - 100\right]$$

$$= p\frac{150}{1 + r} + \frac{50}{1 + r} - 100.$$

Setting this equal to zero yields that

$$p = \frac{1 + 2r}{3}.$$

Therefore, the only probability vector $(p, 1 - p)$ that results in a zero expected return for the wager of purchasing the stock has $p = (1+2r)/3$.

In addition, the present value return from purchasing one share of the option is

$$\text{return} = \begin{cases} 50(1 + r)^{-1} - c & \text{if the price is 200 at time 1,} \\ -c & \text{if the price is 50 at time 1.} \end{cases}$$

Hence, when $p = (1 + 2r)/3$, the expected return of purchasing one option share is

$$E[\text{return}] = \frac{1 + 2r}{3} \frac{50}{1 + r} - c.$$

Thus, it follows from the arbitrage theorem that the only value of $c$ for which there will not be a sure win is

$$c = \frac{1 + 2r}{3} \frac{50}{1 + r},$$

that is, when

$$c = \frac{50 + 100r}{3(1 + r)},$$

which is in accord with the result of Section 4.3.1.    □

The absence of arbitrage does not usually result in a unique value for the cost of an option in a one-period problem. Indeed, it does only when we suppose that there are only two possible specified values of the option at the exercise time. This is illustrated by our next example.

**Example 4.4c**   Consider Example 4.4b, but now suppose that the present value of the price at time 1 can be any one of the values 50, 200, and 100. That is, we now allow for the possibility that the present value of the price of the stock at time 1 is unchanged from its initial price. Again, suppose that we want to price an option to purchase the stock at time 1 for the fixed price of 150. For simplicity, let the interest rate $r$ equal zero. The arbitrage theorem states that there will be no guaranteed win if there are nonnegative numbers $p_{50}, p_{100}, p_{200}$ that sum to 1 and are such that the expected gains if one purchases the stock (or the option) are zero when $p_i$ is the probability that the stock's price at time 1 is $i$ for $i = 50, 100, 200$. Letting $G_s$ denote the gain at time 1 from buying one share of the stock and letting $S(1)$ be the present value of the price of the stock at time 1, we have that

$$G_s = \begin{cases} 100 & \text{if } S(1) = 200, \\ 0 & \text{if } S(1) = 100, \\ -50 & \text{if } S(1) = 50. \end{cases}$$

Hence,

$$E[G_s] = 100p_{200} - 50p_{50}.$$

Also, if $c$ is the cost of the option then the gain from purchasing one option is

$$G_o = \begin{cases} 50 - c & \text{if } S(1) = 200, \\ -c & \text{if } S(1) = 100 \text{ or } S(1) = 50. \end{cases}$$

Therefore,

$$E[G_o] = (50 - c)p_{200} - c(p_{50} + p_{100})$$
$$= 50p_{200} - c.$$

Equating both $E[G_s]$ and $E[G_o]$ to zero shows that the conditions for the absence of arbitrage are that there be probabilities and a cost $c$ such that

$$p_{200} = \frac{1}{2}p_{50} \quad \text{and} \quad c = 50p_{200}.$$

The first of these two equalities implies that $p_{200} \leq 1/3$, so it follows that, for any value of $c$ satisfying $0 \leq c \leq 50/3$, we can find probabilities that make both buying the stock and buying the option fair bets. Therefore, no arbitrage is possible for any option cost in the interval $[0, 50/3]$.    □

## 4.5    The Multiperiod Binomial Model

Let us now consider a stock option scenario in which there are $n$ periods and where the nominal interest rate is $r$ per period. Let $S(0)$ be the initial price of the stock, and for $i = 1, \ldots, n$ let $S(i)$ be its price at $i$ time periods later. Suppose that $S(i)$ is either $uS(i - 1)$ or $dS(i - 1)$, where $d < 1 + r < u$. That is, going from one time period to the next, the price either goes up by the factor $u$ or down by the factor $d$. Suppose also that, at time 0, an option may be purchased that enables one to buy the stock after $n$ periods have passed for the amount $K$. In addition, the stock may be purchased and sold anytime within these $n$ time periods.

Let $X_i$ equal 1 if the stock's price goes up by the factor $u$ from period $i - 1$ to $i$, and let it equal 0 if that price goes down by the factor $d$. That is,

$$X_i = \begin{cases} 1 & \text{if } S(i) = uS(i - 1), \\ 0 & \text{if } S(i) = dS(i - 1). \end{cases}$$

The outcome of the experiment can now be regarded as the value of the vector $(X_1, X_2, \ldots, X_n)$. From the arbitrage theorem it follows that, in order for there not to be an arbitrage opportunity, there must be probabilities on these outcomes that make all bets fair. That is, there must be a set of probabilities

$$P\{X_1 = x_1, \ldots, X_n = x_n\}, \quad x_i = 0, 1, \ i = 1, \ldots, n,$$

that make all bets fair. Now consider the following type of bet. First choose a value of $i$ ($i = 1, \ldots, n$) and a vector $(x_1, \ldots, x_{i-1})$ of 0s and 1s, and then observe the first $i - 1$ changes. If $X_j = x_j$ for each $j = 1, \ldots, i - 1$, immediately buy one unit of stock and then sell it back the next period. If the stock is purchased, then its cost at time $i - 1$ is $S(i - 1)$; the time-$(i - 1)$ value of the amount obtained when it is subsequently sold at time $i$ is either $(1 + r)^{-1}uS(i - 1)$ if the stock goes up or $(1 + r)^{-1}dS(i - 1)$ if it goes down. Therefore, if we let

$$\alpha = P\{X_1 = x_1, \ldots, X_{i-1} = x_{i-1}\}$$

denote the probability that the stock is purchased and let

$$p = P\{X_i = 1 \mid X_1 = x_1, \ldots, X_{i-1} = x_{i-1}\},$$

then the expected gain on this bet (in time-$(i - 1)$ units) is

$$\alpha[p(1 + r)^{-1}uS(i - 1) + (1 - p)(1 + r)^{-1}dS(i - 1) - S(i - 1)].$$

Consequently, the expected gain on this bet will be zero, provided that

$$\frac{pu}{1 + r} + \frac{(1 - p)d}{1 + r} = 1$$

or, equivalently, that

$$p = \frac{1 \mid r \quad d}{u - d}.$$

In other words, the only probability vector that results in an expected gain of zero for this type of bet has

$$P\{X_i = 1 \mid X_1 = x_1, \ldots, X_{i-1} = x_{i-1}\} = \frac{1 + r - d}{u - d}.$$

Since $x_1, \ldots, x_n$ are arbitrary, this implies that the only probability vector on the set of outcomes that results in all these bets being fair is the one that takes $X_1, \ldots, X_n$ to be independent random variables with

$$P\{X_i = 1\} = p = 1 - P\{X_i = 0\}, \quad i = 1, \ldots, n, \qquad (4.2)$$

where

$$p = \frac{1 + r - d}{u - d}.$$

Given these probabilities, it can be shown that any bet on buying stock will have zero expected gain. Hence it follows from the arbitrage theorem that either the cost of the option must equal the expectation of the present (i.e., the time-0) value of owning it under the foregoing probabilities, or else there will be an arbitrage opportunity. So assume that the $X_i$ are independent 0-or-1 random variables with a common probability $p$ of being equal to 1, and note that this implies that their sum – call it $Y$, equal to the number of the $X_i$ that are equal to 1 – is a binomial random variable with parameters $n$ and $p$. Now, in going from period to period, the stock's price is its old price multiplied by either $u$ or by $d$. As a result, $S_n$, the stock's price after $n$ periods, can be expressed as

$$S(n) = u^Y d^{n-Y} S(0),$$

where $Y = \sum_{i=1}^{n} X_i$ is a binomial random variable with parameters $n$ and $p$. The value of owning the option after $n$ periods have elapsed is $(S_n - K)^+$, which is defined to equal $S_n - K$ when this quantity is non-negative or zero when it is negative. Therefore, the present value of the option is

$$(1 + r)^{-n}(S(n) - K)^+$$

and so the expectation of the present value of owning the option is

$$(1+r)^{-n}E[(S(n) - K)^+]$$

$$= (1+r)^{-n}E[(S(0)u^Y d^{n-Y} - K)^+]$$

$$= (1+r)^{-n}\sum_{i=0}^{n}\binom{n}{i}p^i(1-p)^{n-i}(S(0)u^i d^{n-i} - K)^+.$$

Thus, the only option cost $C$ that does not result in an arbitrage is

$$C = (1+r)^{-n}\sum_{i=j}^{n}\binom{n}{i}p^i(1-p)^{n-i}(S(0)u^i d^{n-i} - K), \qquad (4.3)$$

where $j$ is the smallest integer satisfying

$$S(0)u^j d^{n-j} > K.$$

Let $\bar{B}_{n,a}(k)$ denote the probability that a binomial random variable with parameters $n$ and $a$ is greater than or equal to $k$; that is,

$$\bar{B}_{n,a}(k) = \sum_{i=k}^{n}\binom{n}{i}a^i(1-a)^{n-i}.$$

If we set

$$s = \frac{up}{up + d(1-p)} = \frac{up}{1+r}$$

then we can write equation (4.3) as

$$C = S(0)\bar{B}_{n,s}(j) - K(1+r)^{-n}\bar{B}_{n,p}(j).$$

Summing up, we have shown the following.

**Proposition 4.5.1** *For the n-period model in which the price of the stock during any period is equal to its price the previous period multiplied either by u or by d, let S(0) be the initial price and let r be the interest rate per period. If C is the cost of an option to purchase the stock after n time periods at a fixed price K, then there is an arbitrage unless*

$$C = S(0)\bar{B}_{n,s}(j) - K(1+r)^{-n}\bar{B}_{n,p}(j), \qquad (4.4)$$

*where*

$$p = \frac{1+r-d}{u-d}, \qquad s = \frac{up}{1+r},$$

$$\bar{B}_{n,a}(k) = \sum_{i=k}^{n} \binom{n}{i} a^i (1-a)^{n-i},$$

*and where*

$$j = \min\{i : S(0)u^i d^{n-i} > K\}.$$

### 4.5.1    *The Black–Scholes Option Pricing Formula*

Suppose now that we desire to price an option to purchase stock at time $t$ for a price $K$ when the price of the stock changes continuously in time. One approach to this problem is first to break up the time interval from 0 to $t$ into $n$ subintervals of lengths $t/n$. If we then suppose that the price of the stock, in going from one subinterval to the next, either increases by the factor $e^{\sigma\sqrt{t/n}}$ or decreases by the factor $e^{-\sigma\sqrt{t/n}}$ for some positive value $\sigma$, then we have the $n$-period model with

$$u = e^{\sigma\sqrt{t/n}}, \qquad d = e^{-\sigma\sqrt{t/n}},$$

and with the one-period interest rate $rt/n$. Let $C(n)$ denote the nonarbitrage cost (as given by equation (4.4)) of the option in this $n$-period problem. If we now let $n$ grow larger and larger, then the $n$-period problem will become closer and closer to the continuous time problem. It turns out that, as we let $n$ increase, $C(n)$ converges to the value $C(\infty)$ given by

$$C(\infty) = S(0)\Phi(\sigma\sqrt{t} + b) - Ke^{-rt}\Phi(b), \tag{4.5}$$

where

$$b = \frac{rt - \sigma^2 t/2 - \log(K/S(0))}{\sigma\sqrt{t}}$$

and where $\Phi(x)$ is the *standard normal distribution function* of statistics, defined by

$$\Phi(x) = \frac{1}{\sqrt{2\pi}} \int_{-\infty}^{x} e^{-y^2/2} \, dy$$

and equal to the probability that a standard normal random variable is less than $x$.

Equation (4.5) is known as the *Black–Scholes option pricing formula*. The limiting process (as we divide the time interval from 0 to $t$ into more and more subintervals) is known as *geometric Brownian motion*, and $\sigma$ is a measure of the volatility of this process.

## 4.6   Exercises

**Exercise 4.1**   What is the effective interest rate when the nominal interest rate of 10% is

(a) compounded semiannually;
(b) compounded quarterly;
(c) compounded continuously?

**Exercise 4.2**   The *doubling rule* states that if one earns interest at a nominal rate of $s$ percent per year compounded annually, then it will take approximately $70/s$ years for your funds to double. Give a justification for this rule, and see how well it works when

(a) $s = 7$;
(b) $s = 10$.

**Exercise 4.3**   If you receive 5% interest compounded yearly, approximately how many years will it take for your money to quadruple? What if you were earning only 4%?

**Exercise 4.4**   How much do you need to invest at the beginning of each of the next 60 months to amass a value of $100,000 at the end of 60 months, given that the annual nominal interest rate will be fixed at 6% and will be compounded monthly?

**Exercise 4.5**   The yearly cash flows of an investment are

$$-1,000, \ -1,200, \ 800, \ 900, \ 800.$$

Is this a worthwhile investment for someone who can both borrow and save money at the yearly interest rate of 6%?

**Exercise 4.6**   Consider two possible sequences of end-of-year returns,

$$20, 20, 20, 15, 10, 5 \quad \text{and} \quad 10, 10, 15, 20, 20, 20.$$

Which sequence is preferable if the interest rate, compounded annually, is (a) 3%; (b) 5%; (c) 10%?

**Exercise 4.7** Suppose it is known that the price of a certain security after one period will be one of the values $s_1, \ldots, s_k$. What should be the cost of an option to purchase the security at time 1 for the price $K$ if $K < \min s_i$?

**Exercise 4.8** Let $C$ be the price of a call option to purchase a security whose present price is $S$. Argue that $C \leq S$.

**Exercise 4.9** Let $P$ be the price of a put option to sell a security, whose present price is $S$, for the amount $K$. Which of the following are true?

(a) $P \leq S$;
(b) $P \leq K$.

**Exercise 4.10** Let $P$ be the price of a put option to sell a security, whose present price is $S$, for the amount $K$. Argue that

$$P \geq Ke^{-rt} - S,$$

where $t$ is the exercise time and $r$ is the interest rate.

**Exercise 4.11** Consider an experiment with three possible outcomes and odds as follows.

| Outcome | Odds |
|:-------:|:----:|
| 1 | 1 |
| 2 | 2 |
| 3 | 5 |

Is there a betting scheme that results in a sure win?

**Exercise 4.12** Consider an experiment with four possible outcomes, and suppose that the quoted odds for the first three of these outcomes are as follows.

| Outcome | Odds |
|:-------:|:----:|
| 1 | 2 |
| 2 | 3 |
| 3 | 4 |

What must be the odds against outcome 4 if there is to be no possible arbitrage when one is allowed to bet both for and against any of the outcomes?

**Exercise 4.13** Repeat Exercise 4.11 when the odds are

| Outcome | Odds |
|---------|------|
| 1 | 2 |
| 2 | 2 |
| 3 | 2 |

**Exercise 4.14** Suppose, in Exercise 4.11, that one is also allowed to choose any pair of outcomes $i \neq j$ and bet that the outcome will be either $i$ or $j$. What should the odds be on these three bets if an arbitrage opportunity is to be avoided?

**Exercise 4.15** In Example 4.4a, show that if

$$\sum_{i=1}^{m} \frac{1}{1+o_i} \neq 1$$

then the betting scheme

$$x_i = \frac{(1+o_i)^{-1}}{1 - \sum_{i=1}^{m}(1+o_i)^{-1}}, \quad i = 1, \ldots, m,$$

will always yield a gain of exactly 1.

**Exercise 4.16** In Example 4.4b, suppose that one also has the option of purchasing a put option that allows its holder to put the stock for sale at the end of one period for a price of 150. Determine the value of $P$, the cost of the put, if there is to be no arbitrage; then show that the resulting call and put prices satisfy the put–call option parity formula.

**Exercise 4.17** Suppose that, in each period, the cost of a security either goes up by a factor of 2 or down by a factor of 1/2 (i.e., $u = 2$ and $d = 1/2$). If the initial price of the security is 100, determine the no-arbitrage cost of a call option to purchase the security at the end of two periods for a price of 150.

# 5. Graphs and Trees

## 5.1 Graphs

A *graph* consists of a set of elements $\mathcal{V}$ called *vertices* (or *nodes*) and a set $\mathcal{A}$ of pairs of distinct vertices called *edges* (or *arcs*). It is usual to represent such a system graphically by drawing circles for vertices and drawing lines between vertices $i$ and $j$ when $(i, j)$ is an edge. For instance, the graph having $\mathcal{V} = \{1, 2, 3, 4, 5, 6\}$ and $\mathcal{A} = \{(1, 2), (1, 4), (1, 5), (2, 3), (2, 5), (3, 5), (5, 6)\}$ is represented in Figure 5.1.

It should be noted the edges have no direction – for instance, the edge $(1, 3)$ can also be written as $(3, 1)$ – and also that we allow neither edges from a vertex to itself nor multiple edges connecting the same pair of vertices.

A sequence of vertices $i, i_1, i_2, \ldots, i_k, j$ for which $(i, i_1), (i_1, i_2), \ldots, (i_{k-1}, i_k), (i_k, j)$ are all edges is called a *path* from vertex $i$ to vertex $j$. Figure 5.2 shows a path from vertex 1 to vertex 6.

A path $i, i_1, i_2, \ldots, i_k, i$ from a vertex back to itself in which all of the edges $(i, i_1), (i_1, i_2), \ldots, (i_{k-1}, i_k), (i_k, i)$ are distinct is called a *cycle*; see Figure 5.3.

We say that vertices $i$ and $j$ *communicate* either if $i = j$ or if there is a path between $i$ and $j$. We use the symbol $i \longleftrightarrow j$ to denote that $i$ and $j$ communicate. The proof of the following is left as an exercise.

**Proposition 5.1.1** *If $i \longleftrightarrow j$ and $j \longleftrightarrow k$ then $i \longleftrightarrow k$.*

If $i$ and $j$ communicate then we say that they are in the same *component*. It is a consequence of Proposition 5.1.1 that any two components are either identical or disjoint. For consider two components and suppose that vertex $i$ is in one of them and $j$ in the other. Now either these components are disjoint or they have at least one vertex in common. But if they do have a vertex (say, $k$) in common then, since $i$ and $j$ both communicate with $k$, it follows from Proposition 5.1.1 that they communicate with each other, implying that the two components are identical. Thus,

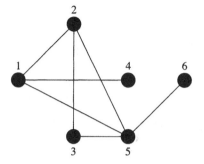

Figure 5.1: A Graph

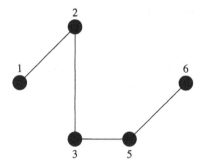

Figure 5.2: A Path from 1 to 6: 1, 2, 3, 5, 6

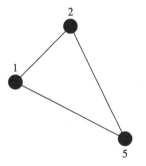

Figure 5.3: A Cycle: 1, 2, 5, 1

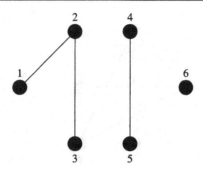

Figure 5.4: A Graph with Three Components

nonidentical components are disjoint, which implies that the vertices of a graph can be partitioned into disjoint components. For instance, Figure 5.4 illustrates a graph consisting of three components.

A graph with a single component is said to be *connected*. That is, a graph is connected if, for all pairs of vertices $i \neq j$, there is a path connecting $i$ and $j$. The graph illustrated by Figure 5.1 is connected whereas the one in Figure 5.4 is not. By convention, a graph with a single vertex is connected.

**Proposition 5.1.2**  *A graph on $n$ vertices that has more than $\binom{n-1}{2}$ edges is connected.*

*Proof.* Consider a graph $G$ with $n$ vertices that is not connected, and let $i$ and $j$ be vertices such that there is no path from $i$ to $j$. Let $S$ denote the set consisting of vertex $i$ and all other vertices for which there is a path from $i$ to that vertex. Note that if $|S|$ denotes the number of vertices in $S$ then, since $i \in S$ and $j \notin S$, $1 \leq |S| \leq n - 1$. Now if $s \in S$ and $r \notin S$ then $(s, r)$ is not an edge of the graph. For if it were then there would be a path from $i$ to $r$ (contradicting the fact that $r \notin S$) – namely, the path that goes from $i$ to $s$ and then into $r$. As a result, if we let $N$ denote the number of pairs of vertices that are not edges of $G$, then

$$N \geq |S|(n - |S|) \geq \min_{1 \leq k \leq n-1} k(n - k) = n - 1.$$

Hence, as there are a total of $\binom{n}{2}$ possible edges, it follows that

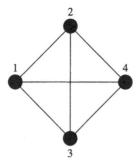

Figure 5.5: The Complete Graph on Four Vertices

$$\text{number of edges of } G = \binom{n}{2} - N$$

$$\leq \binom{n}{2} - (n-1)$$

$$= (n-1)\left(\frac{n}{2} - 1\right)$$

$$= \binom{n-1}{2}. \qquad \square$$

The graph with vertex set $\mathcal{V} = \{1, \ldots, n\}$ whose edges are all of the $\binom{n}{2}$ pairs $(i, j)$, $i \neq j$, is called the *complete graph* on $\mathcal{V}$. Figure 5.5 shows the complete graph on the vertices $\{1, 2, 3, 4\}$.

The *degree* of vertex $i$ is the number of edges of type $(i, j)$, that is, the number of edges that connect $i$ with another vertex. For instance, in the complete graph of $n$ vertices, each vertex has degree $n - 1$. A vertex having degree 1 is called a *leaf*. Let $d(i)$ be the degree of vertex $i$. The graph of Figure 5.1 has two leaves, vertices 4 and 6; the degrees of the other vertices are $d(1) = 3$, $d(2) = 3$, $d(3) = 2$, and $d(5) = 4$.

## 5.2    Trees

A *tree* is a connected graph without any cycles. Figure 5.6 presents trees having one, two, three, four, and five vertices.

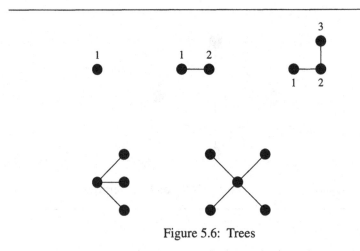

Figure 5.6: Trees

The following theorem is fairly apparent.

**Theorem 5.2.1**   *A graph with n vertices is a tree if and only if it has n − 1 edges and no cycles.*

*Proof.* Suppose first that the graph is a tree, and note that if we remove an edge from the tree then, since a tree has no cycles, there will no longer be a path between the vertices of the removed edge. Hence, removing an edge from a tree results in a graph having two components, each of which is without a cycle. If we now remove a second edge then, by the same reasoning, what remains is a graph with three components, each of which is without a cycle. Continuing, we see that after we have removed n − 1 edges, what remains is a graph with n components; that is, one without any more edges.

To go the other way, suppose we have a set of n − 1 edges that does not contain any cycles. Starting with a graph consisting of n components – that is, consisting of the n vertices and no edges – add these n − 1 edges one at a time. Since each added edge must be between vertices in different components (for otherwise it would result in a cycle), it follows that each added edge decreases the number of components by one. Thus,

Figure 5.7: All Trees When $n = 2$ and $n = 3$

after the $(n - 1)$th edge has been added, the graph has only one component and no cycles; in other words, it is a tree.    □

Recall that the degree of a vertex is the number of edges of which it is part,
and let $D$ be the sum of the degrees of all the vertices of a graph. Each additional edge of a graph adds two to the count of $D$ (since it increases the
degrees of both of its endpoint vertices), so it follows that $D$ is equal to
twice the number of edges of the graph. Hence, $D = 2(n - 1)$ when the
graph is a tree. Since every vertex in a tree with $n > 1$ vertices has a positive degree (since the tree is connected), at least two of the vertices must
be leafs. Otherwise, there would be $n - 1$ or more nonleafs and, since the
degree of each nonleaf is at least 2, the sum of the degrees would thus exceed $2(n - 1)$, which is not the case. Thus, we have shown the following.

**Proposition 5.2.1**  *Every tree has at least two leaves.*

Let us now consider the problem of determining how many trees there
are when $V = \{1, 2, \ldots, n\}$. Figure 5.7 depicts all the trees when $n = 2$
and $n = 3$. The following result gives the number of distinct trees.

**Proposition 5.2.2**  (Cayley's Theorem)  *There are $n^{n-2}$ trees on a vertex set of size $n$.*

To prove Cayley's theorem we first use Corollary 2.7.2 to obtain the following combinatorial identity.

**Lemma 5.2.1**

$$n^{n-2} = \sum_{j=0}^{n-1} \binom{n}{j} j^{n-2} (-1)^{n-j+1}.$$

**Proof.** Corollary 2.7.2 implies that

$$\sum_{j=0}^{n}\binom{n}{j}j^{n-2}(-1)^{n-j+1} = 0,$$

which is equivalent to the identity stated in the lemma.              □

**Proof of Cayley's Theorem.** For a set $B$, let $N(B)$ denote the number of elements of $B$. Also, let $t(n)$ denote the number of trees on a set of $n$ vertices. Our proof is by induction on $n$. As the result is true when $n = 1$ and $n = 2$, assume that it is true whenever the vertex set is of size smaller than $n$. Now consider the vertex set $V = \{1, \ldots, n\}$, where $n > 2$. Let $L_i$ denote the set of trees on $V$ for which vertex $i$ is a leaf, $i = 1, \ldots, n$. Since every tree has at least one leaf, we obtain from the inclusion–exclusion rule that

$$t(n) = N(L_1 \cup L_2 \cup \cdots \cup L_n)$$

$$= \sum_i N(L_i) - \sum\sum_{i<j} N(L_i L_j) + \cdots$$

$$+ (-1)^{n+1} N(L_1 \cdots L_n). \tag{5.1}$$

A tree with vertex $i$ as a leaf can be thought of as consisting of a (sub)tree on the other $n-1$ vertices plus an edge that connects vertex $i$ to any of the $n-1$ vertices in the (sub)tree. Hence, it follows that

$$N(L_i) = (n-1)t(n-1).$$

Similarly, a tree in which $i$ and $j$ are both leaves is equivalent to a tree on the other $n-2$ vertices plus two additional edges, one connecting $i$ to any of the $n-2$ vertices and the other connecting $j$ to any them. Thus,

$$N(L_i L_j) = (n-2)^2 t(n-2).$$

Indeed, the same argument gives the general result that the number of trees for which each of $k$ specified vertices is a leaf is $(n-k)^k t(n-k)$. Note that the first summation in equation (5.1) is over $n$ terms, the second is over $\binom{n}{2}$ terms, the third over $\binom{n}{3}$ terms, and so on. From the preceding we therefore have that

$$t(n) = \sum_{k=1}^{n-1} \binom{n}{k}(n-k)^k t(n-k)(-1)^{k+1}.$$

Using the induction hypothesis, we obtain that

$$\begin{aligned}
t(n) &= \sum_{k=1}^{n-1} \binom{n}{k}(n-k)^k(n-k)^{n-k-2}(-1)^{k+1}\\
&= \sum_{k=1}^{n} \binom{n}{k}(n-k)^{n-2}(-1)^{k+1} \quad \text{(since } n > 2)\\
&= \sum_{j=0}^{n-1} \binom{n}{n-j} j^{n-2}(-1)^{n-j+1} \quad \text{(letting } j = n - k)\\
&= \sum_{j=0}^{n-1} \binom{n}{j} j^{n-2}(-1)^{n-j+1}\\
&= n^{n-2},
\end{aligned}$$

where the final equality follows from Lemma 5.2.1.            $\square$

## 5.3    The Minimum Spanning Tree Problem

Suppose that we are to construct a communications network among $n$ locations and that the cost of building a direct link between locations $i$ and $j$ is $c(i, j) > 0$. The requirements on the network are that enough links must be constructed so that any pair of locations can communicate, possibly through intermediary locations. In other words, regarding the links built as being edges of a graph connecting the $n$ locations (vertices), we desire to construct the cheapest connected graph. Since all costs are positive, it makes no sense to allow any cycles; thus, the cheapest connected graph will be a tree. The problem of finding the cheapest such tree is known as the *minimum spanning tree* problem.

The *greedy algorithm* for the minimum spanning tree problem constructs the edges in sequence as follows: at each stage, build the cheapest edge that – when added to the already constructed graph – does not result in a cycle.

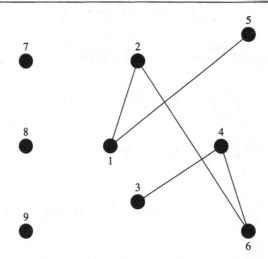

Figure 5.8: Adding $(2, 5)$, the Cheapest Edge, Would Result in a Cycle

**Example 5.3a**   Suppose that the costs for constructing a link between vertices $i$ and $j$ are as follows:

|   | 1 | 2 | 3 | 4 | 5 | 6 | 7 | 8 | 9 |
|---|---|---|---|---|---|---|---|---|---|
| 1 |   | 3.6 | 4.8 | 5.2 | 2.5 | 3.9 | 6.2 | 7.0 | 5.4 |
| 2 |   |   | 8.1 | 7.6 | 3.8 | 3.7 | 5.1 | 4.4 | 6.2 |
| 3 |   |   |   | 2.4 | 6.0 | 5.8 | 7.7 | 9.2 | 10 |
| 4 |   |   |   |   | 8.0 | 3.7 | 6.5 | 7.0 | 6.9 |
| 5 |   |   |   |   |   | 4.0 | 5.5 | 6.6 | 7.7 |
| 6 |   |   |   |   |   |   | 8.1 | 8.8 | 5.9 |
| 7 |   |   |   |   |   |   |   | 5.7 | 5.6 |
| 8 |   |   |   |   |   |   |   |   | 7.3 |

The greedy algorithm starts by building the cheapest edge, namely $(3, 4)$ at cost 2.4; then the next cheapest, namely $(1, 5)$ at cost 2.5; then $(1, 2)$ at cost 3.6; then $(4, 6)$ at cost 3.7 (choosing among same-priced edges is arbitrary); and then $(2, 6)$ at cost 3.7. The next cheapest edge is $(2,5)$; however, since its addition would result in a cycle (see Figure 5.8), it is not included and instead we go on to the next cheapest, and so on until a tree is obtained (see Figure 5.9 for the resulting tree).    □

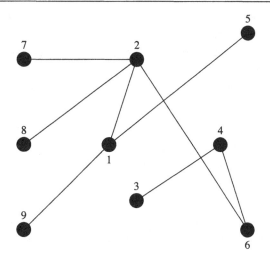

Figure 5.9: The Minimal Cost Spanning Tree

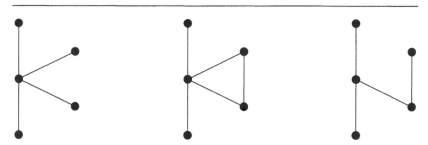

Figure 5.10: Adding an Edge to a Tree; Then Removing an Edge
from the Resulting Cycle

We will show that the greedy algorithm is optimal. However, before
doing so, let us note that if we add a new edge to a tree then the resulting graph contains a cycle; and if we subsequently delete any edge of
this cycle, another tree results (see Figure 5.10).

**Proposition 5.3.1**  *The greedy algorithm results in a minimal cost spanning tree.*

***Proof.*** Let $g_1, g_2, \ldots, g_{n-1}$ denote a possible sequence of greedy edges in order of appearance. (If edge costs are not all distinct then there are different possible greedy edge sequences.) Consider any other spanning tree $T$; we will show that the cost of the greedy tree is less than or equal to that of $T$. In order to show this, let $g_k$ denote the first greedy edge that is not in $T$. That is, $T$ contains $g_1, \ldots, g_{k-1}$ but not $g_k$. Add edge $g_k$ to the tree $T$; this results in a cycle that contains at least one nongreedy edge (since the greedy edges do not contain a cycle). If we now remove a nongreedy edge of the cycle, what remains is a tree that contains the first $k$ greedy edges. However, since $g_k$ is the cheapest edge that does not form a cycle when added to the first $k - 1$ greedy edges, it follows that $g_k$ is at least as cheap as the nongreedy edge that was removed (since that edge and the first $k - 1$ greedy edges are in $T$ and so do not form a cycle). Hence the new tree, which has the first $k$ greedy edges, is at least as cheap as $T$. Repeating this argument shows that the tree composed of the $n - 1$ greedy edges is at least as cheap as $T$; since $T$ was arbitrary, this completes the proof.  □

## 5.4    Cliques and Independent Sets

A set of $k$ vertices of a graph is called a *clique* of size $k$ if all the $\binom{k}{2}$ unordered vertex pairs from this set are edges of the graph. For instance, the vertices $1, 2, 3, 4$ of the graph depicted in Figure 5.11 constitute a clique of size 4. A clique of size 3 is called a *triangle*. For instance, any three of the vertices $1, 2, 3, 4$ of the graph depicted in Figure 5.11 form a clique of size 3.

The following proposition gives a sufficient condition for a graph to contain a triangle.

**Proposition 5.4.1** *A graph with $2m$ vertices and $m^2 + 1$ edges necessarily contains a triangle.*

***Proof.*** The proof is by induction on $m$. It is true when $m = 1$ because a graph with two vertices cannot have two edges. So assume that a graph with $2m$ vertices and $m^2 + 1$ edges necessarily contains a triangle, and consider a graph of $2(m + 1)$ vertices and $(m + 1)^2 + 1$ edges. Select an edge of this graph, say $(i, j)$, and let $N$ denote the number of edges of the graph that do not connect to either $i$ or $j$. There are two cases:

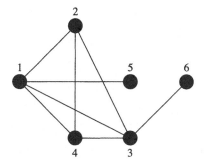

Figure 5.11: A Graph with a Clique of Size 4

(1) $N > m^2$;

(2) $N \leq m^2$.

In case (1), we have that there are more than $m^2$ edges among the $2m$ vertices not equal to $i$ or $j$. By the induction hypothesis this implies that there is a triangle among these vertices. In case (2), since the graph has $(m + 1)^2 + 1$ edges, it follows that

$$\text{number of edges involving either } i \text{ or } j = (m + 1)^2 + 1 - N$$
$$\geq (m + 1)^2 + 1 - m^2$$
$$= 2m + 2.$$

Upon subtracting the edge $(i, j)$, we see in case (2) that at least $2m + 1$ edges go from either $i$ or $j$ to one of the other $2m$ vertices. But this implies that at least one of the other $2m$ vertices is connected to both $i$ and $j$, thus showing that there is also a triangle in case (2). $\qquad\square$

To show that the critical number $m^2 + 1$ in Proposition 5.4.1 cannot be improved upon, consider a graph consisting of $2m$ vertices and $m^2$ edges that is obtained by (i) partitioning the $2m$ vertices into two groups of size $m$ each and then (ii) having edges between every pair of vertices that are not in the same group and no edges between vertices in the same group. Since every set of three vertices contains at least two that are in the same group, it follows that those two do not have an edge and so the three vertices do not form a triangle (see Figure 5.12).

Figure 5.12: A Graph with $2m$ Vertices, $m^2$ Edges, and No Triangles ($m = 3$)

***Remark.*** Our proof of Proposition 5.4.1 made use of the *pigeonhole principle* introduced in Section 2.8, which states that if $r$ objects are to be placed in $s$ pigeonholes then, when $r > s$, at least one pigeonhole will contain more than one object. This self-evident result can be extremely useful in graph theory, as indicated by our next example.

**Example 5.4a**    In a graph with $n$ vertices, show that at least two of the vertices have the same degree.

***Solution.*** There are two cases: either all of the vertices have a positive degree, or there is a vertex having degree 0. In the former case, the possible degrees of the $n$ vertices are $1, \ldots, n - 1$, which implies by the pigeonhole principle that at least two of the vertices have the same degree. (Imagine that the vertices are the objects and that each vertex is to be put in pigeonhole number $i$, $i = 1, \ldots, n - 1$, if it has degree $i$.) In the latter case there is an isolated vertex (the one with degree 0) and so the possible degrees of the $n$ vertices are $0, 1, \ldots, n - 2$. Thus, there are again only $n - 1$ possible degrees for the $n$ vertices and so the pigeonhole principle yields that at least two vertices will have the same degree. □

If $G$ is a graph then its *complement* graph, denoted $G^c$, is the graph whose vertices are the vertices of $G$ but where $(i, j)$ is an edge of $G^c$ if and only if it is *not* an edge of $G$. Figure 5.13 presents a graph and its complement. It should be noted that the complement of the complement is the original graph.

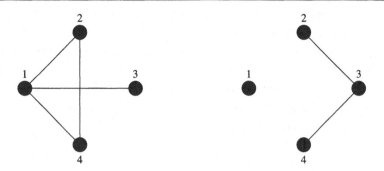

Figure 5.13: A Graph and Its Complement

A set of vertices $S$ of a graph is said to be an *independent set* if no pair of vertices in $S$ constitute an edge. Cliques and independent sets are related through the complement graph concept in that $S$ is a clique of $G$ if and only if $S$ is an independent set of $G^c$.

The number of vertices in the largest independent set is called the *independence number* of the graph and is denoted by $\alpha(G)$. Let $A_i$ denote the set of all vertices adjacent to $i$; that is,

$$A_i = \{j : (i, j) \text{ is an edge of } G\}.$$

The number of elements in $A_i$ is $d(i)$, the degree of vertex $i$. The following proposition gives a lower bound on the independence number of a graph in terms of the degrees of its vertices.

**Proposition 5.4.2**  *For a graph having n vertices,*

$$\alpha(G) \geq \sum_{i=1}^{n} \frac{1}{d(i)+1}.$$

In proving this proposition we will use a technique known as the *probabilistic method,* which involves the introduction of randomness into a nonrandom problem so as to enable the use of results from probability theory.

*Proof.* Let $\mathbf{R} = (R_1, \ldots, R_n)$ denote a random permutation of the numbers $1, 2, \ldots, n$; that is, $\mathbf{R}$ is equally likely to be any of the $n!$

permutations. Now define an independent set of vertices $S$ as follows. Let $i \in S$ if, in the permutation $\mathbf{R}$, $i$ appears before any of the elements in $A_i$. To show that $S$ is an independent set we note that, if $i \in S$ and $(i, j)$ is an edge, then $i$ appears before $j$ in the permutation $\mathbf{R}$ and so $j \notin S$ (since $i \in A_j$ and $j$ does not appear before $i$). Thus, $S$ is an independent set. Let $|S|$ denote the number of elements in $S$. Since $E[|S|]$ is a weighted average of the possible values of $|S|$, each of which is less than or equal to $\alpha(G)$, it follows that

$$\alpha(G) \geq E[|S|].$$

However, setting

$$I_i = \begin{cases} 1 & \text{if } i \in S, \\ 0 & \text{if } i \notin S, \end{cases}$$

we have

$$|S| = \sum_{i=1}^{n} I_i.$$

Using the result that the expected value of the sum of random variables is equal to the sum of the expected values, the preceding equation yields

$$E[|S|] = \sum_{i=1}^{n} E[I_i]$$

$$= \sum_{i=1}^{n} P\{i \in S\}$$

$$= \sum_{i=1}^{n} \frac{1}{d(i) + 1},$$

where the final equality follows because each of the $d(i) + 1$ values, consisting of $i$ and the $d(i)$ values in $A_i$, are equally likely to appear first in the permutation $\mathbf{R}$, implying that the probability that $i$ appears before any of the vertices in $A_i$ is $1/(d(i) + 1)$. Since $\alpha(G) \geq E[|S|]$, the result is proven. $\qquad\square$

Let $G$ be a graph with $n$ vertices, and let $d^c(i)$ be the degree of vertex $i$ in $G^c$. Then, since every pair $(i, j)$, $j \neq i$, is an edge in either $G$ or $G^c$ (but not both), it follows that

$$d^c(i) + d(i) = n - 1.$$

Thus, from Proposition 5.4.2 we obtain that

$$\alpha(G^c) \geq \sum_{i=1}^{n} \frac{1}{n - d(i)}.$$

Since an independent set of $G^c$ corresponds to a clique in $G$, the preceding inequality yields the following proposition.

**Proposition 5.4.3** *If $G$ is a graph with $n$ vertices, then*

$$maximal\ clique\ size\ in\ G \geq \sum_{i=1}^{n} \frac{1}{n - d(i)}.$$

The following corollary is of some independent interest.

**Corollary 5.4.1** *If $G$ has $n$ vertices and $a$ edges, then*

$$maximal\ clique\ size\ in\ G \geq \frac{n^2}{n^2 - 2a}.$$

**Proof.** Since the sum of the vertex degrees is twice the number of edges, the hypothesis implies that

$$\sum_{i=1}^{n} d(i) = 2a.$$

Hence, using Proposition 5.4.3, we have that

maximum clique size

$$\geq \sum_{i=1}^{n} \frac{1}{n - d(i)}$$

$$\geq \min\left\{ \sum_{i=1}^{n} \frac{1}{n - x_i} : \sum_{i=1}^{n} x_i = 2a,\ 0 \leq x_i < n \right\}. \quad (5.2)$$

We may use the convexity of the function $f(x) = \frac{1}{n-x}$ $(0 \leq x < n)$ along with the continuous analog of Corollary 1.4.1 to show that the minimal value of (5.2) occurs when all $x_i = 2a/n$. This proves the result.    □

Corollary 5.4.1 implies the following generalization of Proposition 5.4.1.

**Corollary 5.4.2** (Turan's Theorem)    *A graph with $m(k-1)$ vertices and $\binom{k-1}{2}m^2 + 1$ edges contains a clique of size $k$.*

*Proof.* With $n = m(k-1)$ and $a = \binom{k-1}{2}m^2 + 1$, it follows that

$$2a = (k-1)(k-2)m^2 + 1 = n^2\frac{k-2}{k-1} + 1 > n^2\frac{k-2}{k-1}.$$

Hence,

$$\frac{n^2}{n^2 - 2a} > \frac{n^2}{n^2 - n^2\frac{k-2}{k-1}} = k - 1,$$

which implies (from Proposition 5.4.1) that there is a clique of size $k$.    □

**Remark.** Turan's theorem is sharp in the sense that a graph with $m(k-1)$ vertices and $\binom{k-1}{2}m^2$ edges need not contain a clique of size $k$. For take $m(k-1)$ vertices that are partitioned into $k-1$ subsets of size $m$ each, and consider a graph having these vertices and having edges between every pair of vertices that are in different subsets. An edge can be chosen by first selecting two of the $k-1$ subsets and then selecting one of the $m$ vertices from each of these subsets, so it follows that there are $\binom{k-1}{2}m^2$ edges. However, since there are only $k-1$ subsets, we have that every set of $k$ vertices will contain at least two that are in the same subset; as these two vertices will not be connected by an edge, the set of $k$ vertices will not be a clique.

The probabilistic method, which we used to prove Proposition 5.4.2, is very useful in graph theory. The following example provides another illustration.

**Example 5.4b**    A round-robin tournament involving $n$ contestants is one in which each of the $\binom{n}{2}$ pairs of contestants play each other exactly once, with the outcome on any play being that one member of the pair wins and the other loses. Let the players be numbered $1, 2, \ldots, n$.

The permutation $j_1, j_2, \ldots, j_n$ is called a *Hamiltonian permutation* if $j_1$ beats $j_2$, $j_2$ beats $j_3, \ldots,$ and $j_{n-1}$ beats $j_n$. A problem of interest is to determine the largest possible number of Hamiltonian permutations.

For instance, if $n = 3$ and if one of the players (say, number 1) beats the other two players and 2 beats 3, then 1, 2, 3 will be the only Hamiltonian permutation. On the other hand, if each player wins once, there will be three Hamiltonian permutations. (For instance, if 1 beats 2, 2 beats 3, and 3 beats 1, then 1, 2, 3 and 2, 3, 1 and 3, 1, 2 are all Hamiltonian permutations.) Hence, for $n = 3$, the largest possible number of Hamiltonian permutations is three. Although there is no simple expression for the number of Hamiltonian permutations in the general case, we will now use the probabilistic method to show that, for any $n$, there is a possible outcome of the tournament for which there are at least $n!/2^{n-1}$ Hamiltonian permutations.

To verify the preceding claim, let us suppose in any game played that each of the two participants is equally likely to win and that the outcomes from different games are independent. If we let $X$ denote the number of Hamiltonian permutations that result, then $X$ is a random variable whose set of possible values is the set of all possible numbers of Hamiltonian permutations that can result from a round-robin tournament of $n$ players. Since $E[X]$ is a weighted average of its set of possible values, it follows that at least one of these possible values is at least as large as $E[X]$. To compute $E[X]$, imagine that we have numbered the $n!$ different permutations of $1, 2, \ldots, n$; call them permutation $1, \ldots,$ permutation $n!$. Now, let

$$I_i = \begin{cases} 1 & \text{if permutation } i \text{ is a Hamiltonian,} \\ 0 & \text{otherwise,} \end{cases}$$

and note that

$$X = \sum_{i=1}^{n!} I_i.$$

As a result,

$$E[X] = \sum_{i=1}^{n!} E[I_i].$$

However,

$$E[I_i] = P\{\text{permutation } i \text{ is a Hamiltonian}\}$$
$$= (1/2)^{n-1},$$

where the final equality is true because the probability that any permutation (say, $j_1, j_2, \ldots, j_n$) is a Hamiltonian permutation is the probability
that $j_1$ beats $j_2$ multiplied by the probability that $j_2$ beats $j_3$, multiplied
by the probability that $j_3$ beats $j_4$, and so on. But since each of these
probabilities is $1/2$, the preceding equation follows. Consequently, we
obtain that

$$E[X] = \frac{n!}{2^{n-1}},$$

which shows that there is at least one outcome of the round-robin tournament that results in at least $n!/2^{n-1}$ Hamiltonian permutations.    □

## 5.5    Euler Graphs

A cycle is said to be an *Euler cycle* (sometimes called an *Euler closed
path*) if it is a cycle that passes through every edge of the graph exactly
once. A graph that possesses an Euler cycle is called an *Euler graph*.
For instance, the graph on the left side of Figure 5.14 is an Euler graph
whereas the one on the right side is not. (Why not?)

The following proposition characterizes Euler graphs.

**Proposition 5.5.1**   *A connected graph with at least one vertex is an
Euler graph if and only if the degree of every vertex is even.*

**Proof.**   Suppose first that the graph has an Euler cycle – say, $i, i_1, i_2, \ldots,$
$i_k, i$. Because every edge of the graph is accounted for in this cycle, we
can count the degree of each vertex by adding the number of edges
that go into the vertex to the number that go out of the vertex. However, every time the cycle enters one of the vertices $i_j$ ($i_j \neq i$) it then
leaves it, adding 2 to its degree count. As a result, the degree of each
of these vertices is even. Similarly, aside from the initial and final
edges, which both add 1 to the degree count of vertex $i$, every other
encounter of $i$ adds 2 to this count. Thus the degree of vertex $i$ is also
even.

Suppose now that the graph is connected and the degree of each vertex is an even number. Starting at any vertex, move along any edge
to the next vertex and put a mark on the edge that you have just tra-
versed. From then on, move along an unmarked edge that emanates

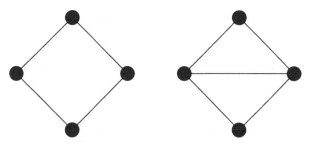

Figure 5.14: An Euler Graph and a Non-Euler Graph

from the vertex you are presently at, and mark that edge. Since there are only a finite number of edges, at some point you will have to stop because there are no unmarked edges connected to the vertex you are at. Consider the vertex where you end. Note that, for any vertex different from the initial one, every time you enter and then leave that vertex you mark two of the edges connected to it; since it has an even number of such edges, it follows that there are still an even number of unmarked edges connected to it. As a result, it is not possible that you could enter one of these vertices and then be unable to leave it. Thus, the final vertex must be the initial one and so we have generated a cycle. If this cycle encompasses all the edges of the graph then we have found an Euler cycle. If not, then the vertices of the graph consisting of the unmarked edges will all have even degrees, so we can start at one of those vertices and repeat the process to obtain a second cycle, and so on.

Thus, whenever each vertex of a graph $G$ has an even degree, we can partition the edges of $G$ into (say, $r$) cycles such that each edge of $G$ is contained in exactly one of these cycles and no cycle contains the same edge more than once. We will now argue by induction that, if such a graph is connected, it must be an Euler graph. As this is true by definition when $r = 1$, assume it is true whenever we can partition the edges into $r - 1$ such cycles, and suppose now that $G$ can be partitioned into $r$ cycles. Let $C_1$ be one of the $r$ cycles. Since $G$ is connected, it follows that at least one of the vertices in $C_1$ must also be a vertex of at least

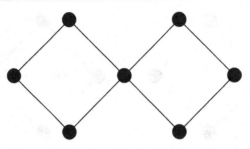

Figure 5.15:  Combining Two Cycles Having a Common Vertex

one of the other cycles. That is, for some vertices $i$, $j$, $k$ there must be a cycle $C_s$ ($s \neq 1$) such that

$$(i, j) \in C_1, \qquad (j, k) \in C_s.$$

But then (see Figure 5.15) $C_1$ and $C_s$ together form a single cycle that passes through each of its edges exactly once, and so $G$ can be represented as a connected graph consisting of $r - 1$ such cycles. This implies (by the induction hypothesis) that $G$ is an Euler graph, and the proof is complete.                                                                □

## 5.6    Exercises

**Exercise 5.1**    Prove Proposition 5.1.1.

**Exercise 5.2**    Is there a graph with five vertices whose respective degrees are 2, 2, 3, 3, 3?

**Exercise 5.3**    If $d(i) \geq 2$ for all vertices $i$, show that the graph contains a cycle.

**Exercise 5.4**    How many distinct graphs having vertex set $\mathcal{V} = \{1, 2, \ldots, n\}$ are possible? How many have exactly $k$ edges?

**Exercise 5.5**    The *distance* between two vertices $i \neq j$, call it $d(i, j)$, is defined to be the length of the shortest $i$-$j$ path in the graph, where the length of a path is defined to be the number of edges it contains. If there is no $i$-$j$ path, define the distance to be infinite. With $d(i, i) = 0$, show that for any vertices $i, j, k$ we have

$$d(i, j) \le d(i, k) + d(k, j).$$

**Exercise 5.6** Consider a collection of sets $\{S_1, \ldots, S_r\}$. Associated with this collection is the *intersection* graph whose vertices are $1, \ldots, r$ and which has $(i, j)$ as an edge, provided that $S_i$ and $S_j$ have a nonempty intersection. Draw the intersection graph if $r = 5$ and

(a) $S_1 = \{a, b, c, d\}$;
(b) $S_2 = \{c, e\}$;
(c) $S_3 = \{a, e, g\}$;
(d) $S_4 = \{a, b, e\}$;
(e) $S_5 = \{d\}$.

**Exercise 5.7** Show that a graph having $n$ vertices, $n - 1$ edges, and no cycles is a tree.

**Exercise 5.8** A system of roads connecting eight different locations must be built. The distances (in miles) between each pair of locations is as follows.

|   | 2 | 3 | 4 | 5 | 6 | 7 | 8 |
|---|---|---|---|---|---|---|---|
| 1 | 13 | 21 | 9 | 7 | 18 | 20 | 15 |
| 2 |   | 9 | 18 | 12 | 26 | 23 | 11 |
| 3 |   |   | 26 | 17 | 25 | 19 | 10 |
| 4 |   |   |   | 7 | 16 | 15 | 9 |
| 5 |   |   |   |   | 9 | 11 | 8 |
| 6 |   |   |   |   |   | 6 | 10 |
| 7 |   |   |   |   |   |   | 5 |

Assuming that the total cost is proportional to the sum of the distances of the roads constructed, what is the cheapest solution?

**Exercise 5.9** A mining company is about to start operations at six mining camps. The company wants to construct a series of roads so that each camp is reachable from every other camp. It the cost of construction is $c$

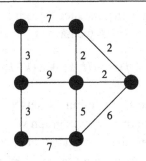

Figure 5.16

dollars per mile of construction and the distances between these camps are as follows, find the optimal solution.

|   | 2 | 3 | 4 | 5 | 6 |
|---|---|---|---|---|---|
| 1 | 5 | 2 | 9 | 12 | 7 |
| 2 |   | 14 | 5 | 8 | 1 |
| 3 |   |   | 4 | 6 | 21 |
| 4 |   |   |   | 17 | 3 |
| 5 |   |   |   |   | 13 |

**Exercise 5.10**    Consider the following approach for finding a minimum cost spanning tree. Start with the complete graph; at each stage, find a cycle and remove the most expensive edge of that cycle; stop when there no longer exists any cycles. Does this approach work?

**Exercise 5.11**    How many distinct minimum spanning trees does the graph in Figure 5.16 possess?

**Exercise 5.12**    Another algorithm for finding a minimal spanning tree is known as *Prim's algorithm*. At each stage of the algorithm, $X$ is a set of vertices and $T$ is the cheapest set of edges that connect vertices in $X$. The final value of $T$ will be the minimal cost spanning tree. The algorithm proceeds as follows:

(1)  Let $X = \{1\}$ and $T = \emptyset$ (i.e., $T$ is the null set).
(2)  Choose the cheapest edge $(i, j)$ such that $i \in X$ and $j \notin X$.

(3) Let $X = X \cup \{j\}$ and $T = T \cup \{(i, j)\}$.

(4) If $X \neq V$, return to step (1).

In words, at each stage the algorithm chooses the cheapest edge that connects a vertex of the edges so chosen with a new vertex. Prove that Prim's algorithm finds a minimal cost spanning tree.

**Exercise 5.13**   Apply Prim's algorithm to the graph of Exercise 5.8.

**Exercise 5.14**   Suppose that one edge of a graph is colored blue and that each of the other edges is colored either red or green. Show that either:

(a) there is a cycle in which no edge is green; or
(b) there is a cycle in which no edge is red.

**Exercise 5.15**   Let $G^c$ be the complement of the graph $G$.

(a) If $G$ has $n$ vertices and $m$ edges, how many edges does $G^c$ have?
(b) Describe the complement of the complete graph.
(c) What is the complement of $G^c$?
(d) If $G$ is connected, show that $G^c$ is not connected.

**Exercise 5.16**   The *diameter* of $G$ is the maximum distance between two vertices in $G$. Show that, if the diameter of $G$ is greater than 3, then the diameter of $G^c$ is less than 3.

**Exercise 5.17**   A path that uses all the edges of a graph – but does not end where it began – is called a *semi-Eulerian* path. Show that a connected graph contains a semi-Eulerian path if and only if it has exactly two vertices of odd degree.

**Exercise 5.18**   If each of the vertices of a graph $G$ can be colored with any of $k$ different colors in such a way that every two adjacent vertices have a different color, we say that the graph is *k-colorable*. (Vertices $i$ and $j$ are said to be "adjacent" if $(i, j)$ is an edge.) The smallest $k$ for which a graph is $k$-colorable is called its *chromatic number,* usually designated as $\chi(G)$.

(a) If $K_n$ is the complete graph on $n$ vertices, what is $\chi(K_n)$?
(b) What is $\chi(T)$ when $T$ is a tree with $n$ vertices?

(c) The *circuit graph* $Z_n$ is the graph with vertex set $\{1, \ldots, n\}$ and edge set $\{(1, 2), (2, 3), \ldots, (n - 1, n), (n, 1)\}$. What is $\chi(Z_n)$?

**Exercise 5.19**   There is no simple algorithm for finding the chromatic number of a graph, but the following (known as the "greedy" algorithm) can be used to obtain an upper bound. To begin, arbitrarily order the vertices of the graph – say the ordering is $v_1, \ldots, v_m$. Then proceed as follows.

- Use color 1 on vertex $v_1$, and set $c(1) = 1$.
- If $(v_1, v_2)$ is not an edge, use color 1 on $v_2$ and set $c(2) = 1$; otherwise use color 2 and set $c(2) = 2$.
- For $i > 1$, use color $r$ on vertex $v_i$ and set $c(i) = r$, where $r$ is the smallest positive integer not in the set $\{c(j) : j < i, (v_i, v_j) \text{ is an edge}\}$.
- Continue until all vertices have been colored, and note how many colors are needed.

(a) Show, by constructing an example along with a vertex order, that the greedy algorithm may give an answer that is larger than the chromatic number.
(b) Show that, for at least one ordering of the vertices, the greedy algorithm gives the chromatic number.

**Exercise 5.20**   Every member of a certain organization belongs to several committees, and a schedule of committee meetings is to be drawn up. Each committee is to meet exactly once, but any two committees with at least one member in common cannot meet at the same time. Show that determining the minimal number of meeting times required is equivalent to finding the chromatic number of a certain graph.

**Exercise 5.21**   What is the relation between the chromatic number and the maximum clique size of a graph?

**Exercise 5.22**   Recall that a set of vertices $W$ of a graph $G$ is said to be an *independent set* if no two of the vertices in the set are adjacent; $\alpha(G)$, the *independence number* of $G$, is defined as the size of a largest independent set. Find the independence number of the graph depicted in Figure 5.15.

**Exercise 5.23**    If the graph $G$ has $n$ vertices, show that

$$\alpha(G) \geq \frac{n}{\chi(G)},$$

where $\chi(G)$ is the chromatic number of $G$.

**Exercise 5.24**    If $d(i) \leq d$ for all vertices $i$, argue that

$$\chi(G) \leq d + 1,$$

where $\chi(G)$ is the chromatic number of the graph $G$.

**Exercise 5.25**    Suppose we have $m$ colors that can be used to color the vertices of a tree with $n > 1$ vertices. Show that, subject to the condition that any vertices joined by an edge must have a different color, there are $m(m - 1)^{n-1}$ different colorings possible.

*Hint:* Use mathematical induction.

# 6. Directed Graphs

## 6.1    Directed Graphs

A graph whose edges are assumed to have a direction is called a *directed graph*, or more simply a *digraph*. The edge $(i, j)$ in a directed graph is interpreted as going from vertex $i$ into vertex $j$, and it is graphically represented by drawing an arrow from vertex $i$ to vertex $j$. Figure 6.1 presents a directed graph.

As with an undirected graph, any sequence of vertices $i, i_1, i_2, \ldots, i_k, j$ for which $(i, i_1), (i_1, i_2), \ldots, (i_k, j)$ are all edges is said to be a *path* from vertex $i$ to vertex $j$. For instance, 1, 2, 3, 6 is a path from 1 to 6 for the digraph of Figure 6.1.

## 6.2    The Maximum Flow Problem

Consider a directed graph with vertex set $V$ and edge set $A$. Suppose that there are two distinguished vertices, the *source* vertex $s$ and the *sink* vertex $t$, and suppose that there are no edges that either go into $s$ or out of $t$. For each edge $(i, j)$, suppose that a nonnegative integer $c(i, j)$, called the "capacity" of the edge, is specified. The problem of interest is to flow a commodity from $s$ to $t$ according to the following rules:

(1)  the amount of flow that can be sent along any edge is less than or equal to the capacity of that edge (the *capacity* constraint);

(2)  for any vertex $i$ not equal to $s$ or $t$, the amount that is flowed into $i$ must equal the amount that is flowed out of $i$ (the *conservation* constraint).

Our objective is to determine the maximal amount that can be sent from $s$ to $t$ and the flow that achieves it.

For the directed graph given in Figure 6.2, the edge numbers on the upper graph represent the capacities; those on the lower graph represent a possible flow along that edge. Note that each edge flow value is less than or equal to the edge capacity and that, except for the source and

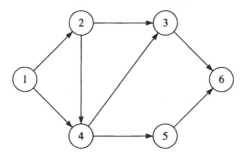

Figure 6.1:  Directed Graph with
$\mathcal{A} = \{(1, 2), (1, 4), (2, 3), (2, 4), (3, 6), (4, 3), (4, 5), (5, 6)\}$

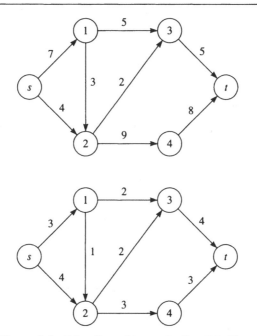

Figure 6.2:  Edge Capacities and a Feasible Flow

sink vertices, the total amount of flow into each vertex is equal to the total flow out of that vertex.

Let us introduce some notation. For sets of vertices $U$ and $V$, let $(U, V)$ denote the set of edges that go from a vertex of $U$ to one of $V$. That is,

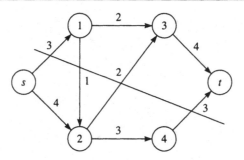

Figure 6.3: An *s-t* Cut with $X = \{s, 2, 4\}$

$$(U, V) = \{(i, j) \in \mathcal{A} : i \in U, \ j \in V\}.$$

Also, for a function $g(i, j)$ defined on the edges of the graph, let

$$g(U, V) = \sum_{(i, j) \in (U, V)} g(i, j)$$

denote the sum of the $g(i, j)$, summed over all the edges that go from a vertex in $U$ to one in $V$.

Using the preceding notation, our objective is to find nonnegative edge flow values $f(i, j)$ so as to

$$\max_f v(f),$$

subject to

$$f(i, j) \leq c(i, j)$$

and

$$f(i, V) = f(V, i), \quad i \neq s, t,$$

where $v(f) = f(s, V) = f(V, t)$ is the amount that $f$ flows from $s$ to $t$.

An important concept in finding a maximal flow is that of an *s-t* cut. Let $X$ be a set of vertices and let $\bar{X}$ be the complementary set of remaining vertices. If $s \in X$ and $t \in \bar{X}$ then we say that $(X, \bar{X})$ is an *s-t cut*. In other words, an *s-t* cut is a set of edges that go from a set of vertices that includes $s$ to the set of remaining vertices that includes $t$. A useful way of thinking about an *s-t* cut is to imagine that a river runs through the network, with the vertices in $X$ on one side of the river and those in $\bar{X}$ on the other side. For instance, Figure 6.3 depicts an *s-t* cut with $X = \{s, 2, 4\}$ for the graph of Figure 6.2. The edge numbers represent a flow.

Since $f(X, \bar{X})$ is the total flow from the $s$ side to the $t$ side of the cut and $f(\bar{X}, X)$ is the total amount that flows back, it follows that the net amount flowing across the cut is $f(X, \bar{X}) - f(\bar{X}, X)$. From the conservation condition, it follows that this must equal the value of the flow (see Figure 6.3). That is, we have the following proposition.

**Proposition 6.2.1** *For any s-t cut $(X, \bar{X})$ and flow $f$,*

$$v(f) = f(X, \bar{X}) - f(\bar{X}, X).$$

As an instance of Proposition 6.2.1 we observe that, for the flow and $s$-$t$ cut indicated in Figure 6.3, the value of the flow is clearly 7 while $f(X, \bar{X}) = 8$ and $f(\bar{X}, X) = 1$.

An immediate corollary of Proposition 6.2.1 is that the value of any flow is less than or equal the capacity of any $s$-$t$ cut.

**Corollary 6.2.1** *For any flow f and s-t cut $(X, \bar{X})$,*

$$v(f) \le c(X, \bar{X}).$$

*Proof.* We have

$$v(f) = f(X, \bar{X}) - f(\bar{X}, X)$$
$$\le f(X, \bar{X})$$
$$\le c(X, \bar{X}).$$

The first inequality follows since $f$ is nonnegative, and the second follows since the flow along an edge is less than or equal to the edge's capacity. $\square$

Corollary 6.2.1 is valid for all flows; hence, for any $s$-$t$ cut $(X, \bar{X})$ we have

$$\max_f v(f) \le c(X, \bar{X}).$$

As this is true for all $s$-$t$ cuts, we are led to our next corollary.

**Corollary 6.2.2**

$$\max_f v(f) \le \min c(X, \bar{X}),$$

*where the minimum is over all s-t cuts $(X, \bar{X})$.*

Suppose now that we can find a flow $f_o$ and an *s-t* cut $(X_o, \bar{X}_o)$ such that $v(f_o) = c(X_o, \bar{X}_o)$. Since the value of any other flow is at most $c(X_o, \bar{X}_o)$, it follows that $f_o$ is the maximal flow and (also from Corollary 6.2.2) that $(X_o, \bar{X}_o)$ is the minimal capacity *s-t* cut. It turns out that this is always possible, resulting in the following theorem.

**Theorem 6.2.1** (Max-Flow Min-Cut Theorem)

$$\max_f v(f) = \min c(X, \bar{X}).$$

*Proof.* We will prove the theorem by presenting an algorithm, known as the *augmentation algorithm*, for solving the maximum flow problem. Starting with any initial integral flow $f$, the algorithm will at each iteration either produce a new integral flow whose value is larger than the preceding one or stop. When it stops, we will be able to determine an *s-t* cut whose capacity is equal to the value of the final flow. Since we will then have exhibited a flow and an *s-t* cut such that the value of the flow is equal to the capacity of the cut, the result will follow.

Let $f$ be any integral flow; for instance, we could start the algorithm with the flow $f(i, j) = 0$, $(i, j) \in A$. Now, define the *augmented digraph* associated with the flow $f$ in the following manner. The vertex set $\mathcal{V}$ is as before, but the edge set $A^f$ now consists of all pairs of vertices $(i, j)$ for which it is possible to increase the amount that $f$ flows from $i$ to $j$ and still preserve the capacity constraint. There are two types of edges in $A^f$:

(1) if $(i, j) \in A$ and $f(i, j) < c(i, j)$, then $(i, j) \in A^f$;
(2) if $(j, i) \in A$ and $f(j, i) > 0$, then $(i, j) \in A^f$.

In the former case we say that $(i, j)$ is a *forward* edge of the augmented digraph and in the latter case that it is a *reverse* edge. Thus, if $f$ sends flow along the edge $(i, j)$ that is less than the capacity of that edge, then we can increase the flow in a direct fashion by sending additional flow along that edge. On the other hand, if $f$ sends positive flow along the edge $(j, i)$ then we can (indirectly, or in a reverse fashion) increase the amount flowed directly from $i$ to $j$ by decreasing the amount that is being sent along $(j, i)$. For instance, ignoring any edges into the source vertex $s$ or out of the sink vertex $t$, the augmented digraph associated with the flow depicted in Figure 6.2 consists of the set of edges shown in Figure 6.4.

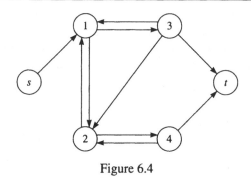

Figure 6.4

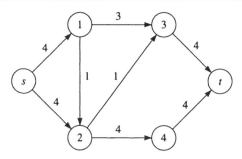

Figure 6.5: An Improved Flow

Suppose now that there is a path from $s$ to $t$ along the edges in $\mathcal{A}^f$. By increasing the flow by 1 along all of the edges in that path, either by a direct increase if the edge is a forward edge or by decreasing the amount sent in the opposite direction if it is a reverse edge, we obtain a new flow $f'$ such that $v(f') = v(f) + 1$. For instance, increasing the flow of Figure 6.2 by sending 1 additional unit of flow along the path $s, 1, 3, 2, 4, t$ results in the flow, of value 8, shown in Figure 6.5. (Note that the effect of sending an additional unit of flow along the reverse edge (3, 2) is to reduce the amount sent along (2, 3) by the amount 1.)

Thus, whenever we can find a path from $s$ to $t$ in the augmented digraph, we are able to obtain a new flow whose value is one more than its predecessor. Since flow values are bounded by the capacity of any $s$-$t$ cut, eventually we will reach a point where we have a flow $f_o$ for which there is no path from $s$ to $t$ along the edges $\mathcal{A}^{f_0}$. We claim that this flow has the maximal possible flow value, and to prove this we will define a cut whose capacity will be shown to equal $v(f_0)$. So suppose there is no path from $s$ to $t$ along the edges $\mathcal{A}^{f_0}$. Let $X_o$ denote the set of vertices consisting of $s$ and all vertices $i$ for which there is a path from $s$ to $i$ in $\mathcal{A}^{f_0}$, and let $\bar{X}_o$ denote the remaining vertices. Since there is no path from $s$ to $t$, it follows that $t \in \bar{X}_o$.

Suppose now that $i \in X_o$ and $j \in \bar{X}_o$. Note that, since there is a path from $s$ to $i$ but not from $s$ to $j$, it follows that $(i, j) \notin \mathcal{A}^{f_0}$. (Otherwise, there would be a path from $s$ to $j$: one that goes from $s$ to $i$ and then goes along the augmented edge $(i, j)$.) Hence, if $(i, j) \in \mathcal{A}$ then

$$f_0(i, j) = c(i, j)$$

and if $(j, i) \in \mathcal{A}$ then

$$f_0(j, i) = 0.$$

From the preceding, it follows that

$$f(X_o, \bar{X}_o) = c(X_o, \bar{X}_o), \qquad f(\bar{X}_o, X_o) = 0.$$

Therefore, from Proposition 6.2.1,

$$v(f_0) = c(X_o, \bar{X}_o),$$

which completes the proof. $\qquad\qquad\qquad\qquad\qquad\qquad\qquad\qquad$ □

**Remark.** When actually utilizing the augmentation algorithm, it is advantageous to define edge capacities of the augmented network corresponding to the flow $f$ as follows:

$$c^f(i, j) = \begin{cases} c(i, j) - f(i, j) & \text{if } (i, j) \text{ is a forward edge of } \mathcal{A}^f, \\ f(j, i) & \text{if } (i, j) \text{ is a reverse edge of } \mathcal{A}^f. \end{cases}$$

In words, $c^f(i, j)$ is the amount by which the flow $f$ can be increased along the edge $(i, j)$ without violating the capacity constraints. Hence, if there is an $s$-$t$ path in the augmented network then, rather than just sending 1 additional unit of flow along this path, we can speed up the algorithm by sending as much flow as possible – namely, the minimal value of $c^f(i, j)$ over all edges $(i, j)$ in the path.

**Example 6.2a**  With the edge numbers representing the additional capacities $c^f(i, j)$, the augmented digraph of the flow given in Figure 6.5

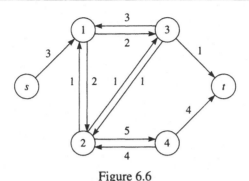

Figure 6.6

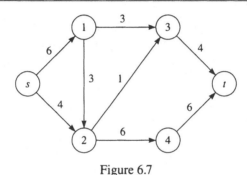

Figure 6.7

is shown in Figure 6.6. (Note that we ignore all reverse edges that either go in to the source vertex $s$ or out of the sink vertex $t$.) From this augmented digraph we see that it is possible to send 2 additional units of flow along the path $s, 1, 2, 4, t$ and so obtain a new flow of value 10, as seen in Figure 6.7. The augmented digraph of this flow is shown in Figure 6.8.

Thus we can send an additional unit of flow along the path $s, 1, 3, t$ to obtain the new flow, of value 11, as shown in Figure 6.9. The new augmented digraph is depicted in Figure 6.10.

There is no path from $s$ to $t$ in the augmented digraph, so we can conclude that the preceding flow, whose value is 11, is optimal. Also, since in the final augmented digraph there is no path from $s$ to any other vertex, it follows that the minimal $s$-$t$ cut is

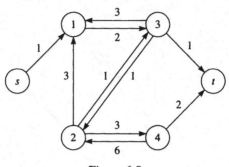

Figure 6.8

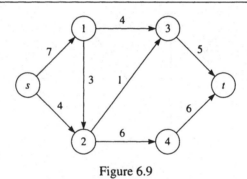

Figure 6.9

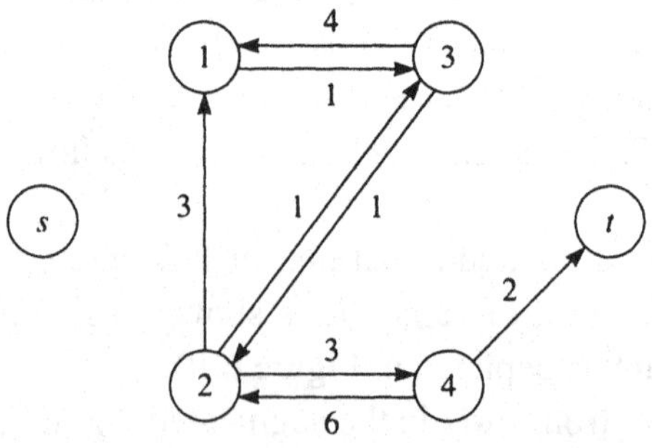

Figure 6.10

$$X = \{s\}, \qquad \bar{X} = \{1, 2, 3, 4, t\},$$

and

$$c(X, \bar{X}) = c(s, 1) + c(s, 2) = 11. \qquad \square$$

***Remark.*** Not only does the augmentation algorithm prove the max-flow min-cut theorem, it also establishes that if the capacities are all integers then the maximal flow can be taken to be integral. That is, there is a maximal flow having only integer flow values $f(i, j)$. When the capacities are rational, then we can write them all as integral multiples of a fixed rational value $r$, and thus the augmentation algorithm would yield an optimal flow with all of its values also being multiples of $r$. Although our argument would need to be modified, Theorem 6.2.1 remains true even when the edge capacities are arbitrary nonnegative real numbers.

A corollary of the proof of the max-flow min-cut theorem is the graph theoretic result known as Menger's theorem.

**Corollary 6.2.3** (Menger's Theorem)  *In a directed graph, the maximum number of edge-disjoint paths from vertex s to vertex t is equal to the minimal number of edges that need to be removed to disconnect s and t.*

***Proof.*** Let all of the edges have flow capacity 1, and use the augmentation algorithm to find an integral maximal flow. Since the flow along each edge will either be 0 or 1, this flow can be decomposed into unit flows along edge-disjoint paths from $s$ to $t$. Also, since any collection of (say, $k$) edge-disjoint paths can be used to obtain a flow of value $k$, it follows that the maximum flow from $s$ to $t$ is equal to the maximal number of edge-disjoint paths from $s$ to $t$. Moreover, any set of edges whose removal disconnects $s$ from $t$ must contain an $s$-$t$ cut. To see why, suppose that a set of edges disconnecting $s$ and $t$ has been removed, and let $X$ denote the vertices that remain connected with $s$ (i.e., $X$ is the set of vertices $x$ for which there is a path from $s$ to $x$); it is now easy to see that all the edges in the $s$-$t$ cut $(X, \bar{X})$ have been removed. Therefore, since the capacity of a cut is equal to the number of edges in that cut, the result follows from the max-flow min-cut theorem. $\qquad \square$

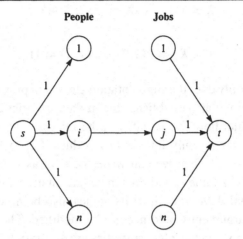

Figure 6.11: The Assignment Problem

## 6.3    Applications of the Maximum Flow Problem

### 6.3.1    *The Assignment Problem*

Consider a set of $n$ people and set of $n$ jobs to which the people are to be assigned. Suppose that at most one person is to be assigned to each job. Suppose further that not every person is qualified for every job, and that the objective is to maximize the number of people that are assigned to jobs for which they are qualified.

This problem can be solved as a maximum flow problem as follows. The digraph has a source vertex, a vertex for each person, a vertex for each job, and a sink vertex. There are edges, each having capacity 1, from the source vertex to each of the people vertices. There is an edge between person vertex $i$ and and job vertex $j$ if person $i$ is qualified for job $j$. Finally, there are edges, each having capacity 1, from each job vertex to the sink vertex. The network is depicted in Figure 6.11. (A digraph along with associated edge numbers is often called a *network*.)

Note that each person vertex can receive only one unit of flow; there are only edges from a person to a job for which that person is qualified; and at most one unit of flow can be sent out of each job. As a result, by regarding a flow that sends one unit from person $i$ to job $j$ as an assignment of person $i$ to job $j$, we see that a flow corresponds to an assignment of people to jobs for which they are qualified. Hence the maximum flow is just the maximum number of people that can be so assigned.

It is interesting to consider conditions under which it is possible to assign everyone. This will be the case when the maximum flow is equal to $n$, so it follows from the max-flow min-cut theorem that this will hold if the capacity of every $s$-$t$ cut is at least $n$. To determine when this is the case, first note that we have not yet assigned any capacities to the edges from people to jobs. It follows that, although it is natural to let these edge capacities equal 1 (since the flow out of a job can be at most 1), any value larger than 1 can also be used for these edge capacities. To make it easier to identify those cuts whose capacities could be minimal, we will suppose that any edge from a person to a job has infinite capacity.

In order to characterize those $s$-$t$ cuts that might have minimal capacity, consider any $s$-$t$ cut and let $X$ denote the vertices on the $s$ side. Let $R$ denote the set of people in $X$, and let $J(R)$ denote the set of jobs that can be performed by at least one member of $R$. Now, if any of the jobs in $J(R)$ are on the $\bar{X}$ side of the cut then the capacity of this cut will be infinite. So suppose that all of the vertices in $J(R)$ are also in $X$. Now consider those jobs not in $J(R)$. Since none of the people in $X$ are qualified for any of those jobs, there are no edges from any of the vertices in $X$ to any of those jobs; hence, the capacity of the cut will be smallest if those jobs are in $\bar{X}$. Thus, a minimal $s$-$t$ cut is of the following type: for some set of people $R$,

$$X = \{s, R, J(R)\}, \qquad \bar{X} = \{\bar{R}, \bar{J}(R), t\},$$

where $\bar{J}(R)$ are all jobs not in $J(R)$. Since $(X, \bar{X})$ consists of edges from $s$ to people in $\bar{R}$ and from jobs in $J(R)$ to $t$, we see that the capacity of a cut of this type is given by

$$c(X, \bar{X}) = |\bar{R}| + |J(R)|$$
$$= n - |R| + |J(R)|,$$

where $|C|$ denotes the number of elements in the set $C$. It follows from the preceding that the capacity of each $s$-$t$ cut will be at least $n$, implying that the maximum flow will equal $n$ if, for every set of people $R$,

$$|J(R)| \geq |R|.$$

That is, we have proven the following result.

**Proposition 6.3.1** (Hall's Theorem)    *It is possible to assign all n people to jobs for which they are qualified if and only if, for every set of people R,*

$$|J(R)| \geq |R|.$$

*That is, the necessary and sufficient condition is that, for every set of people, the number of jobs that at least one member of that set is qualified to perform must be at least as large as the number of people in the set.*

***Remarks.*** (i) The necessity of the condition is obvious. For instance, if there were a set of five people that together qualified for only four jobs, then clearly they could not all be assigned. What is interesting is that the condition is sufficient.

(ii) The way to check whether there is a perfect assignment is to utilize the maximum flow algorithm presented in Section 6.2 to see whether there is a flow of value $n$ (and not by trying to check the condition $|J(R)| \geq |R|$ for each of the $2^n - 1$ possible nonempty sets of people).

**Example 6.3.1a**    Consider the following game of solitaire played with a standard deck of 52 cards (4 suits of 13 different denominations). The cards are turned face up in a rectangular array of 4 rows, each row consisting of 13 columns.

$$
\begin{array}{ccccccccccccc}
X & X & X & X & X & X & X & X & X & X & X & X & X \\
X & X & X & X & X & X & X & X & X & X & X & X & X \\
X & X & X & X & X & X & X & X & X & X & X & X & X \\
X & X & X & X & X & X & X & X & X & X & X & X & X \\
\end{array}
$$

To win this game, you must select one card from each column so that all 13 denominations appear among the 13 cards selected. Argue that, no matter what the arrangement of cards, a win is always possible.

***Solution.*** Consider the problem of assigning 13 people to 13 jobs, and say that person $i$ is qualified for job $j$ if column $i$ contains a card having denomination $j$. By considering the denomination of the card chosen from column $i$ in the solitaire game as being the job assignment of person $i$, it follows that a win in solitaire corresponds to an assignment of all 13 people to jobs for which they are qualified. By Hall's theorem, such an assignment is possible provided that, for any set of $k$ people, the number of jobs that at least one of these $k$ is qualified to do is at least $k$. But since each person is qualified for four jobs and any set of $4k$ cards must contain at least one of each of $k$ different denominations, the result follows.   □

## 6.3.2   *The Tournament Win Problem*

Consider a set of $r$ teams involved in a tournament, and suppose that teams $i$ and $j$ are to play a total of $n_{i,j}$ games between themselves, $i \neq j$. Each game played results in one of the teams being declared the winner. The question of interest is if, for a given integral vector $\mathbf{w} = (w_1, \ldots, w_r)$, it is possible that team $i$ ends up with $w_i$ wins for each $i = 1, \ldots, r$. Since the total number of games to be played is $\frac{1}{2} \sum_j \sum_i n_{i,j}$ and there is one winner in each game, an obvious necessary condition is that

$$ W = \frac{1}{2} \sum_j \sum_i n_{i,j}, $$

where $W = \sum_{i=1}^r w_i$. We will suppose from here on that the preceding condition holds.

We can analyze the tournament as a maximum flow problem by (i) defining vertices for each pair of teams and also for each individual team and (ii) letting the flow sent from a team-pair vertex $(i, j)$ to a team in the pair (say, team $i$) be the number of the $n_{i,j}$ games played by the pair that are won by team $i$. More specifically, consider a network with:

(1) a source vertex $s$;
(2) $\binom{r}{2}$ vertices $(i, j)$, $i < j$;
(3) $r$ vertices $i$, $i = 1, \ldots, r$;
(4) edges from $s$ to each vertex $(i, j)$ with respective capacities $n_{i,j}$;
(5) edges from each vertex of type $(i, j)$ to $i$ and to $j$; and
(6) edges from each vertex $i$ to the source vertex $t$ with respective capacities $w_i$.

The network is depicted in Figure 6.12.

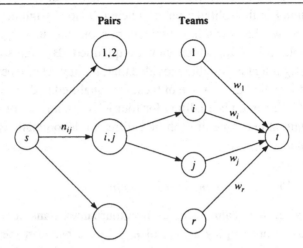

Figure 6.12: Network for the Tournament Win Problem

The problem is to determine if there is an integer flow of value $W = \sum_{i=1}^{r} w_i$. Thus, we would solve for the maximum flow and, if its value is $W$, then $\mathbf{w}$ is a possible win vector. It is interesting to see what the max-flow min-cut theorem implies about when there is a flow of value $W$. To make it easier to identify $s$-$t$ cuts that may be minimal, let us suppose that the capacities of the edges from vertices of the form $(i, j)$ to $i$ (or to $j$) is infinite. That is, although the flow from $(i, j)$ to $j$ is at most $n_{i,j}$, we will allow the capacity of that edge to be $\infty$.

To characterize those $s$-$t$ cuts that might have minimal capacity, consider any $s$-$t$ cut and let $X$ denote the vertices on the $s$ side. Let $T$ denote the set of teams in $\bar{X}$. Consider any team pair vertex involving one of the teams in $T$. If that vertex were in $X$ then the capacity of the cut would be $\infty$; thus, let us suppose that all these vertices are in $\bar{X}$. Consider now a team-pair vertex $(i, j)$, where neither $i$ nor $j$ is in $T$, and note that putting this vertex in $X$ does not add anything to the capacity of the cut. Thus, if we let

$$P(T) = \{(i, j) : i < j, i \in T \text{ or } j \in T\}$$

denote the set of team-pair vertices that involve at least one of the teams in $T$, then it follows from the preceding that a minimal $s$-$t$ cut is of the following type: For some set of teams $T$,

$$X = \{s, \bar{T}, (i, j) \notin P(T)\}, \qquad \bar{X} = \{t, T, (i, j) \in P(T)\}.$$

Now, $(X, \bar{X})$ consists of edges from $s$ to team-pair vertices for which at least one team is in $T$ and from teams that are not in $T$ to $t$. Thus, the capacity of a cut of this type is

$$
\begin{aligned}
c(X, \bar{X}) &= \sum_{(i,j)\in P(T)} n_{i,j} + \sum_{i \notin T} w_i \\
&= \sum_{(i,j)\in P(T)} n_{i,j} + W - \sum_{i \in T} w_i.
\end{aligned}
$$

The capacity of every $s$-$t$ cut is therefore at least $W$ if, for every set of teams $T$,

$$\sum_{(i,j)\in P(T)} n_{i,j} \geq \sum_{i \in T} w_i.$$

That is, we have shown the following.

**Proposition 6.3.2**  *The necessary and sufficient conditions for* **w** *to be a possible win vector are that*

$$\sum_{i=1}^{r} w_i = \frac{1}{2} \sum_j \sum_i n_{i,j}$$

*and, for every set of teams T,*

$$\sum_{(i,j)\in P(T)} n_{i,j} \geq \sum_{i \in T} w_i.$$

*That is, for every set of teams, the number of games involving at least one of the teams in the set must be at least as large as the total number of wins proposed for the teams in the set.*

**Remark.**  As in the case of the assignment problem, the necessity part of this proposition is obvious. For instance, if teams 1 and 2 were involved in a total of 10 games then the result $w_1 = w_2 = 6$ is clearly not possible. Also, for a specified vector **w**, we would check whether it is a possible win vector by using the augmentation algorithm to determine if the maximum flow of the network of Figure 6.12 is equal to $\sum_i w_i$.

### 6.3.3    *The Transshipment Problem*

Suppose that the vertices of a directed graph $G$ with edge capacities are partitioned into two subsets, $S$ and $\bar{S} = C$. The vertices in $S$ are the supplier vertices and those in $C$ are the consumer vertices of a certain commodity. The vertex $i \in S$ desires to supply the commodity at a rate of $a(i)$ per unit time; the vertex $j \in C$ desires to receive it at a rate of $b(j)$ per unit time. The transshipment problem is to determine if, given the edge capacities, it is possible to send flow through the network so as to satisfy the requirements of all the suppliers and all the consumers. An obvious necessary condition is that the total rate at which suppliers desire to supply the commodity must equal the total rate at which consumers desire to receive it; that is, we must have

$$a(S) = b(C)$$

where, if $X$ is a set of vertices and $g$ a function on vertices, $g(X) = \sum_{i \in X} g(i)$.

In order to solve the preceding, known as the *transshipment problem,* we set up a network flow problem as follows:

- adjoin to the original network a source vertex $s$ and a sink vertex $t$;
- add edges from $s$ to each vertex $i \in S$ with edge capacities $a(i)$, respectively;
- add edges from each vertex $j \in C$ to $t$ with respective edge capacities $b(j)$;
- otherwise, leave the network as is.

In the context of this network, the problem is to determine if there is a flow that saturates all of the edges out of $s$ and all those in to $t$; in other words, the problem is to determine if the maximum flow value is equal to $a(S)$. This can be determined for any specified values of the parameters by using the augmentation algorithm.

As always, it is of interest to interpret the max-flow min-cut theorem in the context of this application. To do so, let $Y$ denote any set of vertices of the original graph $G$, let $\bar{Y}$ denote the remaining vertices in $G$, and consider the $s$-$t$ cut

$$X = \{s, \bar{Y}\}, \qquad \bar{X} = \{t, Y\};$$

its capacity is

$$c(X, \bar{X}) = a(S \cap Y) + b(C \cap \bar{Y}) + c(\bar{Y}, Y).$$

Hence, assuming that $a(S) = b(C)$, it follows from the max-flow min-cut theorem that all requirements can be satisfied if, for all sets of vertices $Y$, the preceding is at least $b(C) = b(C \cap \bar{Y}) + b(C \cap Y)$. We have thus shown the following.

**Proposition 6.3.3**   *All requirements of the transshipment problem can be met if and only if*:

(i) $a(S) = b(C)$; *and*
(ii) *for every set of vertices Y,*

$$a(S \cap Y) + c(\bar{Y}, Y) \geq b(C \cap Y).$$

The interpretation of condition (ii) in this proposition is that, for all sets of vertices $Y$, the supply in $Y$ plus the maximal capacity for sending the commodity to $Y$ from outside must be at least as large as the demand within $Y$.

## 6.3.4    An Equipment Selection Problem

A spaceship is setting off to a distant planet and we must decide which of $m$ pieces of equipment $\{1, \ldots, m\}$ should be put on board, where the cost of taking equipment $i$ is the nonnegative value $c_i$, $i = 1, \ldots, m$. The equipment is necessary to perform certain experiments; specifically, there are $n$ experiments, with the $j$th one requiring the set of equipment $S_j \subset \{1, \ldots, m\}$. If experiment $j$ is performed, a nonnegative return $R_j$ is earned. The problem is determining which pieces of equipment should be put on board the ship so as to maximize the sum of the returns obtained for performing experiments minus the sum of the carrying costs of the equipment.

Thus, the problem is to choose $S \subset \{1, 2, \ldots, m\}$ to

$$\text{maximize} \sum_{j:S_j \subset S} R_j - \sum_{i \in S} c_i. \tag{6.1}$$

Noting that

$$\sum_{j:S_j \subset S} R_j = \sum_{j} R_j - \sum_{j:S_j \not\subset S} R_j,$$

it follows that the problem is equivalent to choosing $S$ to

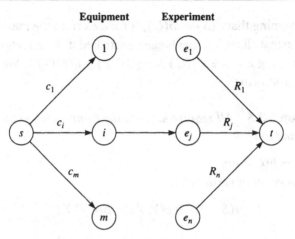

Figure 6.13: The Maximum Flow Digraph for the Equipment Selection Problem

$$\text{minimize} \quad \sum_{j \,:\, S_j \not\subseteq S} R_j + \sum_{i \in S} c_i. \tag{6.2}$$

In other words, the maximum of (6.1) is $\sum_j R_j$ minus the minimum in (6.2). Thus, the original problem can be solved by choosing a set of equipment $S$ so as to minimize the cost of equipment $\left(\sum_{i \in S} c_i\right)$ plus the lost benefit of those experiments that cannot be performed $\left(\sum_{j \,:\, S_j \not\subseteq S} R_j\right)$.

We now show how the preceding can be solved by setting up a maximum flow problem in such a way that the capacity of the minimal $s$-$t$ cut is precisely the quantity (6.2). This is accomplished by letting the digraph have a source vertex $s$, a vertex for each of the $m$ pieces of equipment, a vertex for each of the $n$ possible experiments, and a sink vertex $t$. There are edges from the source vertex to each of the equipment vertices, with the capacity of the edge into equipment vertex $i$ being equal to $c_i$, and there are edges from equipment vertex $i$ to each experiment vertex that requires equipment $i$ for each $i = 1, \ldots, m$. Finally, for each $j = 1, \ldots, n$ there is an edge from experiment vertex $e_j$ to the sink vertex with capacity $R_j$. The digraph is shown in Figure 6.13.

It remains to decide on the appropriate capacities of those edges going from pieces of equipment to experiments that utilize them. Noting that the desired value of an $s$-$t$ cut is of the form $\sum_{i \in S} c_i + \sum_{j:S_j \not\subseteq S} R_j$, it is apparent that we do not want a minimal cut to contain any of edges from equipment to experiments. Let us therefore give those edges a capacity of $\infty$. Let $(X, \bar{X})$ be an $s$-$t$ cut whose capacity might be minimal, and let $\bar{S}$ be the set of equipment on the $s$ side of the cut. Consider any experiment that requires a piece of equipment in $\bar{S}$; if this experiment were on the $t$ side of the cut, then the cut would include an edge from a piece of equipment to an experiment requiring it and so would have infinite capacity. Therefore, all experiments that require any piece of equipment in $\bar{S}$ must also be on the $s$ side of the cut.

Now consider any experiment, say $j$, that does not require any equipment from $\bar{S}$. Putting experiment $j$ on the $s$ side of the cut adds the amount $R_j$ to the capacity of the cut; on the other hand, putting $j$ on the $t$ side of the cut does not add anything to the cut capacity. Therefore, the cuts that might have minimal capacity are of the following type: for some set of equipment $S$,

$$X = \{s, \bar{S}, \text{ all experiments requiring any equipment in } \bar{S}\},$$

$$\bar{X} = \{t, S, \text{ all experiments not requiring any equipment in } \bar{S}\}.$$

Since the capacity of such a cut is

$$c(X, \bar{X}) = \sum_{i \in S} c_i + \sum_{j:S_j \not\subseteq S} R_j,$$

it follows that the equipment selection problem reduces to finding the minimal $s$-$t$ cut of the digraph of Figure 6.13. This can be accomplished by using the augmentation algorithm to find the maximum $s$-$t$ flow and the resulting minimal cut. The equipment nodes on the $t$ side of the minimal $s$-$t$ cut are the ones that should be taken along.

**Example 6.3.4a**   Suppose there are five pieces of equipment and four potential experiments. The requirements and returns from the experiments are as follows.

| $j$ | $S_j$ | $R_j$ |
|---|---|---|
| 1 | {1, 2} | 3 |
| 2 | {2, 3} | 10 |
| 3 | {2, 3, 4} | 6 |
| 4 | {4, 5} | 8 |

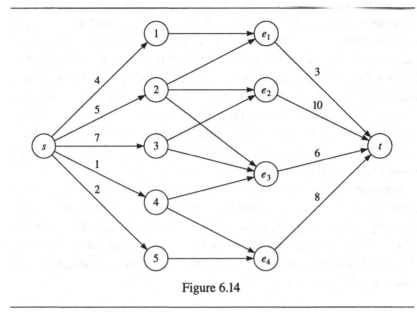

Figure 6.14

The costs of bringing the pieces of equipment are

$$c_1 = 4, \quad c_2 = 5, \quad c_3 = 7, \quad c_4 = 1, \quad c_5 = 2.$$

With the unlabeled edges having capacity $\infty$, the network for the resulting maximum flow problem is depicted in Figure 6.14. It is easy to
verify that the maximal flow has value 18 and that the minimal $s$-$t$ cut
has $X = \{s, 1, e_1\}$ (i.e., the $s$ side of the minimal cut consists of vertex $s$, equipment 1, and experiment 1). Thus, it is optimal to take along
equipment 2, 3, 4, 5; all experiments but the first can be performed, and
the net profit is 9. □

## 6.4    Shortest Path in Digraphs

Suppose that for each edge $(i, j)$ of a directed graph there is a nonnegative number $d(i, j)$ that we sometimes interpret as the distance of
the edge $(i, j)$ and sometimes as the cost of traveling along the edge
$(i, j)$. Starting from a specified vertex $s$, we desire to find the cheapest paths from vertex $s$ to each of the other vertices in the graph, where
the cost of a path is the sum of the costs of the edges in this path.

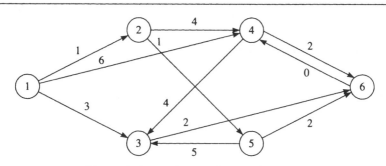

Figure 6.15: A Directed Graph with Costs

For instance, in the directed graph depicted in Figure 6.15, interpreting the numbers on the edges to represent the edge distances yields that the length of the path 1, 2, 5, 3, 6 that goes from vertex 1 to vertex 6 is $d(1, 2) + d(2, 5) + d(5, 3) + d(3, 6) = 9$.

We will now present an algorithm, known as the *Dijkstra algorithm,* for finding the shortest paths from vertex $s$ to each of the other vertices in the directed graph. At every step in the algorithm, each vertex $i$ will have a label $d(i)$ that is either temporary (subject to future change) or permanent. We will put *bars* over the labels to indicate when they are permanent (e.g., $\bar{d}(i)$ indicates that the label on vertex $i$ is permanent). The labels will be shown to have the following interpretations.

- Permanent label: $\bar{d}(i)$ is the length of the shortest path from $s$ to $i$.
- Temporary label: $d(i)$ is the length of the shortest path from $s$ to $i$ that passes only through vertices that have permanent labels.

In addition, vertices are permanently labeled in order of their proximity to $s$. That is, the first vertex with a permanent label will be the closest vertex to $s$, the second vertex having a permanent label will be the second closest vertex to $s$, and so on.

As we describe the algorithm, we will demonstrate its application to the graph of Figure 6.15 (with $s = 1$).

*Step 1.* Let $d(s) = 0$ and $d(i) = \infty$ for $i \neq s$ (see Figure 6.16).

*Step 2.* Choose the vertex having the smallest temporary label – say it is vertex $i$ – and make its label $d(i)$ permanent. For each temporarily labeled vertex $j$ for which $(i, j) \in \mathcal{A}$, reset $d(j)$ by

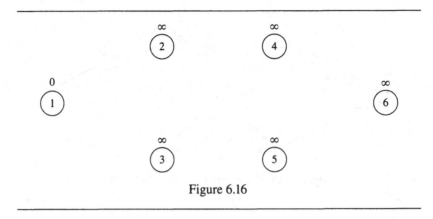

Figure 6.16

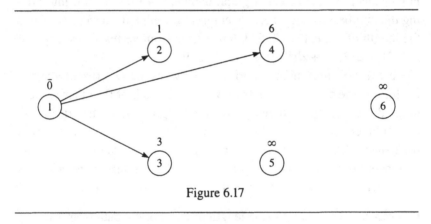

Figure 6.17

$$d(j) = \min\{d(j),\ \bar{d}(i) + d(i, j)\}$$

(see Figure 6.17). In this equation, $d(j)$ on the RHS is its old value and $d(j)$ on the LHS is its reset value. Also, if the value of $d(j)$ is decreased, then add edge $(i, j)$ and remove any other edge leading in to vertex $j$.

*Step 3.* If all the vertices have permanent labels, stop; otherwise, return to step 2 (see Figure 6.18).

**Theorem 6.4.1**    *The Dijkstra algorithm produces the shortest (minimal-cost) paths from vertex s to each of the other vertices.*

*Proof.* We will prove the result by showing, at each stage of the algorithm, the validity of the following:

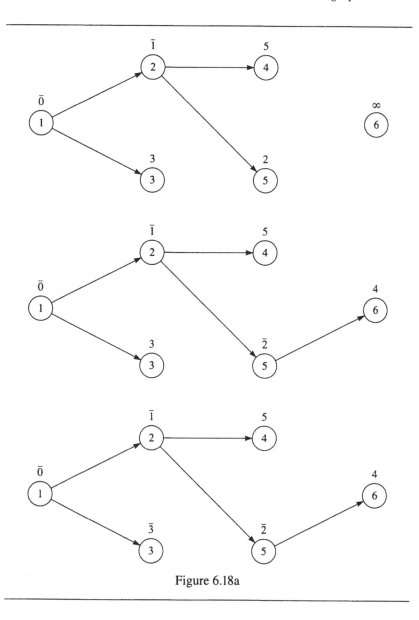

Figure 6.18a

(a) $\bar{d}(i)$ is the length of the shortest path from $s$ to $i$;
(b) the vertices are permanently labeled in order of their proximity to $s$;
(c) $d(i)$ is the length of the shortest path from $s$ to $i$ that passes only through vertices with permanent labels.

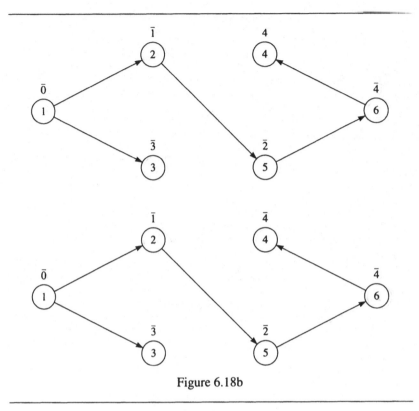

Figure 6.18b

As (a), (b), and (c) are certainly true at the initial step (where $s$ is given a permanent label of 0), assume that they hold when there are $k$ permanently labeled vertices – namely, $s$ and its $k-1$ nearest vertices. Note that the minimal distance path from $s$ to its $k$th nearest vertex (call it vertex $i$) passes only through the $k-1$ permanently labeled vertices. For if it passes through some temporarily labeled vertex (say, vertex $j$) then there would be a path from $s$ to $j$ that is shorter than the minimal distance path from $s$ to $i$, contradicting the assumption that $i$ is the $k$th nearest vertex. It follows that, if vertex $i$ has the smallest temporary label, then $i$ is the $k$th nearest vertex to $s$ and $d(i)$ is the shortest distance from $s$ to $i$. It remains only to show that, when the distance label on $i$ is made permanent, the reset value of $d(j)$ represents the minimum distance from $s$ to $j$ using only $i$ and the other permanently labeled vertices as intermediates. By the induction hypothesis this is clearly the case if this minimum distance path does not go through vertex $i$. On the other

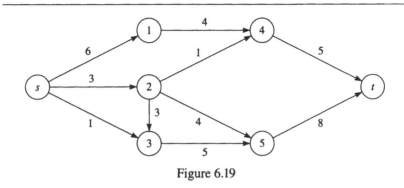

Figure 6.19

hand, if the path does go through $i$, then $i$ must be the vertex leading into $j$; for if the minimum path from $s$ to $j$ using only permanently labeled vertices uses an edge from $i$ to some other permanently labeled vertex, then the minimal distance from $s$ to $i$ would be strictly less than the minimal distance from $s$ to that other vertex, contradicting the fact that $i$ is the farthest from $s$ of all the permanently labeled vertices. Hence, the new value of the minimum distance from $s$ to $j$ (using only $i$ and the other permanently labeled vertices) is as given, and the induction is complete.

To determine the minimum distance paths, define the values $v(j)$, $j \neq s$, as follows. If vertex $i$ has just been given a permanent label and this results in a new value for $d(j)$, reset $v(j)$ to equal $i$. In other words, at each stage of the algorithm, the shortest path from $s$ to $j$ using only the permanently labeled vertices has the final edge $(v(j), j)$. When the algorithm stops, the shortest path from $s$ to $j$ is obtained by noting that its final edge is $(v(j), j)$, the edge preceding that is $(v(v(j)), v(j))$, and so on until the edge $s$ appears. □

## 6.5 Exercises

**Exercise 6.1** Explain how to solve the maximum flow problem when there are vertex constraints, that is, when there are numbers $m_i$ ($i \in \mathcal{V}$) such that the maximal flow into vertex $i$ cannot exceed $m_i$.

**Exercise 6.2** Find the maximum $s$-$t$ flow and the $s$-$t$ cut having minimal capacity for the digraph whose edge capacities are as given in Figure 6.19.

**Exercise 6.3**  Find the maximum $s$-$t$ flow and the $s$-$t$ cut having minimal capacity for a digraph whose edge capacities are as follows:

$$c(s, 1) = 10, \quad c(s, 2) = 15, \quad c(s, 3) = 20,$$
$$c(1, t) = 10, \quad c(1, 4) = 5, \quad c(2, 4) = 12,$$
$$c(3, 4) = 15, \quad c(3, t) = 5, \quad c(4, t) = 20.$$

**Exercise 6.4**  Find the maximal possible flow from vertex 1 to vertex 6 and the minimal cut for a digraph with vertices 1, ..., 6 and with edge capacities as follows:

$$c(1, 2) = 2, \quad c(1, 3) = 1, \quad c(2, 4) = 3, \quad c(2, 5) = 2,$$
$$c(3, 2) = 2, \quad c(3, 5) = 6, , \quad c(3, 4) = 5,$$
$$c(4, 3) = 3, \quad c(4, 6) = 2, \quad c(5, 6) = 4.$$

**Exercise 6.5**  There are four types of disks: tops (t), bottoms (b), covers (c), and centers (e). A disk pack consists of 1 t, 1 b, 1 c, and 9 e. Substitutions can be made as follows:

(a)  e may serve as e, t, b, or c;
(b)  t may serve as either t or c;
(c)  b may serve as either b or c;
(d)  c may serve only as c.

If we have 422 tops, 534 bottoms, 175 covers, and 3,979 centers, how many disk packs can be assembled?

**Exercise 6.6**  Consider three digraphs having the same vertices and edges. Let $c_k(i, j)$ denote the capacity of edge $(i, j)$, and let $V_k$ denote the value of the maximal $s$-$t$ flow in digraph $k$ ($k = 1, 2, 3$). If

$$c_3(i, j) = c_1(i, j) + c_2(i, j),$$

give a relationship between $V_3$ and $V_1 + V_2$.

**Exercise 6.7**  Consider the simple assignment problem with $n$ people and $n$ jobs.

(a)  Show that, if (for some $k > 0$) each person can do exactly $k$ jobs and each job can be done by exactly $k$ people, then there is a way of assigning each person to a job for which she is qualified.

(b) Show that, if each person can do at least half the jobs and each job can be done by at least half the people, then there is a way of assigning each person to a job for which she is qualified.

**Exercise 6.8** In the following matrix, the rows represent teen-age boys and the columns teen-age girls; a 1 indicates that a particular boy–girl pair is allowed to date.

$$\begin{matrix}
0 & 0 & 0 & 1 & 0 & 1 \\
0 & 1 & 0 & 0 & 1 & 0 \\
1 & 0 & 0 & 0 & 0 & 1 \\
0 & 1 & 1 & 0 & 1 & 0 \\
0 & 0 & 0 & 1 & 0 & 1 \\
1 & 0 & 0 & 1 & 0 & 1
\end{matrix}$$

Determine the maximum number of dates that can occur on a given evening.

**Exercise 6.9** An organization of $n$ people wants to set up $m$ standing committees, with the $j$th one to consist of $b_j$ people, $j = 1, \ldots, m$. Suppose person $i$ is qualified for all but is only willing to serve on at most $a_i$ committees, $i = 1, \ldots, n$. Use the max-flow min-cut theorem to derive simple necessary and sufficient conditions for there to exist a possible assignment (of people to committees) that satisfies all constraints.

**Exercise 6.10** A square matrix is called *doubly stochastic* if (i) all its entries are nonnegative and (ii) the row and column sums all equal 1. Prove that an $n \times n$ doubly stochastic matrix contains a set of $n$ nonzero elements, no two of which are in the same row or column.
  *Hint:* Relate this to the assignment problem.

**Exercise 6.11** Consider the maximum flow problem when the flows have lower bounds. That is, suppose that the capacity constraint is replaced by

$$b(i, j) \le f(i, j) \le c(i, j),$$

where the $b(i, j)$ are nonnegative integers. Suppose that we have found a feasible integral flow (i.e., an integral flow that satisfies the preceding as well as the conservation constraint).

(a) Explain how we can modify the augmentation algorithm to find a maximum flow.
(b) Show that the final flow produced by the algorithm is the maximal flow.

*Hint:* Show that, if $f$ is the final flow, then there is an *s-t* cut $X, \bar{X}$ such that

$$f(X, \bar{X}) = c(X, \bar{X}), \qquad f(\bar{X}, X) = b(\bar{X}, X).$$

**Exercise 6.12**   Consider a directed graph without any cycles. Explain how you can label the vertices so that there are no edges $(i, j)$ when $j < i$.

**Exercise 6.13**   The complete graph in which each edge is given a direction is called a *tournament*. By interpreting the edge $(i, j)$ as indicating that player $i$ defeated player $j$, such a graph can be used to model the result of a round-robin tournament. How many distinct tournament graphs on $n$ vertices are possible?

**Exercise 6.14**   In a tournament graph without any cycles, show that there is a unique path that visits every vertex.

**Exercise 6.15**   The equations that follow give the distances of roads connecting various cities. Find the shortest paths, using these roads, from city 1 to each of the other cities.

$$d(1, 2) = 1, \quad d(1, 3) = 2,$$
$$d(2, 3) = 1, \quad d(2, 4) = 5, \quad d(2, 5) = 2,$$
$$d(3, 4) = 2, \quad d(3, 5) = 1, \quad d(3, 6) = 4,$$
$$d(4, 5) = 3, \quad d(4, 6) = 6, \quad d(4, 7) = 8,$$
$$d(5, 6) = 3, \quad d(5, 7) = 7,$$
$$d(6, 7) = 5, \quad d(6, 8) = 2, \quad d(7, 8) = 6.$$

**Exercise 6.16**   For Figure 6.20, use Dijkstra's algorithm to find the shortest path from vertex 1 to all other vertices.

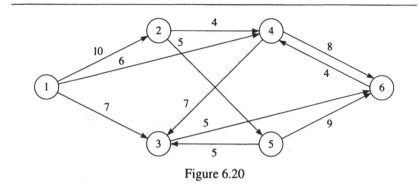

Figure 6.20

**Exercise 6.17**   Consider a digraph with vertices $1, 2, \ldots, n$ and edge costs $d(i, j)$. Let $a_1, \ldots, a_n$ be arbitrary numbers, and set

$$\bar{d}(i, j) = d(i, j) + a_i - a_j.$$

Show that the cheapest path from vertex 1 to vertex $n$ using the edge costs $d(i, j)$ is also the cheapest when using the edge costs $\bar{d}(i, j)$.

# 7. Linear Programming

## 7.1    The Standard Linear Programming Problem

Let us begin this chapter with a pair of examples.

**Example 7.1a**    The following investment opportunity is available from Monday to Friday of the forthcoming week. For any amount $x$, if you invest $x$ on a day and $2x$ on the next day then you will receive $4x$ at the beginning of the third day. The amount received at the beginning of a day can then be used that day for starting a new investment or for continuing an ongoing investment. There are no payments received after Friday (so one should not begin a new investment on Thursday or Friday). If you invest $x$ but cannot put down $2x$ on the next day then you lose the initial investment $x$. For instance, starting with a fortune of 1, it does not pay to invest more than $1/3$ on day 1. For if $x > 1/3$ is invested on day 1 then your remaining capital of $1 - x$ will not be sufficient to enable you to invest $2x$ on day 2, resulting in a forfeit of the initial investment $x$. On the other hand, you could invest $1/3$ on day 1 and $2/3$ on day 2, which yields the return $4/3$ at the the beginning of day 3; you could then use that amount to begin a second investment by investing $4/9$ (one third of $4/3$) on day 3 and $8/9$ on day 4, thus earning a return of $16/9$ on day 5. Can you do better? What is the maximal amount you can have by the end of the week?

*Solution.* You can do better than $16/9$. Indeed, consider the investment stategy given by the following table, in which each row indicates the beginning of a new investment. The numbers without parentheses indicate the amount invested, and the ones in parentheses are the amounts received.

| Monday | Tuesday | Wednesday | Thursday | Friday |
|--------|---------|-----------|----------|--------|
| 1/4    | 1/2     | (1)       |          |        |
|        | 1/4     | 1/2       | (1)      |        |
|        |         | 1/2       | 1        | (2)    |

That is, on Monday one starts an investment of 1/4. On Tuesday, 1/2 is invested to continue Monday's investment, and 1/4 is used to start a new investment. On Wednesday, 1 is collected from Monday's investment; 1/2 of it is then used to fund the investment begun on Tuesday and 1/2 to begin a new investment. On Thursday, 1 is collected from the investment begun on Tuesday, and this is then used to fund the investment begun on Wednesday. On Friday, 2 is collected from the investment begun on Wednesday.

Thus, it is possible to end the week with 2. As a prelude to determining if we can do even better, let us give a mathematical formulation. To do so, let $x_i \geq 0$ denote the amount of the new investment that is to be started on day $i$, $i = 1, 2, 3$. The following table, with the final row indicating the daily amounts of money that remain in savings, describes the investment flows.

| Monday | Tuesday | Wednesday | Thursday | Friday |
|---|---|---|---|---|
| $x_1$ | $2x_1$ | $(4x_1)$ | | |
| | $x_2$ | $2x_2$ | $(4x_2)$ | |
| | | $x_3$ | $2x_3$ | $(4x_3)$ |
| $1 - x_1$ | $1 - 3x_1 - x_2$ | $1 + x_1 - 3x_2 - x_3$ | $1 + x_1 + x_2 - 3x_3$ | $1 + x_1 + x_2 + x_3$ |

Note that such an investment scheme is possible, or *feasible,* provided that the savings at the end of each day is nonnegative. Also, provided that the investment scheme $x_1, x_2, x_3$ is feasible, it results in the final fortune $1 + x_1 + x_2 + x_3$, which is intuitive since an initial investment of size $x$ that is continued until payoff yields the profit $4x - x - 2x = x$. Thus, the problem is to choose $x_1, x_2, x_3$ so as to

$$\text{maximize } 1 + x_1 + x_2 + x_3$$

$$\textit{subject to}$$

$$1 - x_1 \geq 0,$$

$$1 - 3x_1 - x_2 \geq 0,$$

$$1 + x_1 - 3x_2 - x_3 \geq 0,$$

$$1 + x_1 + x_2 - 3x_3 \geq 0;$$

$$x_1 \geq 0, \quad x_2 \geq 0, \quad x_3 \geq 0. \qquad \square$$

Table 7.1: *Percentage of Nutrients in the*
*Different Food Types*

| Nutrient | Food 1 | Food 2 | Food 3 | Food 4 |
|---|---|---|---|---|
| Protein | 14 | 52 | 26 | 6 |
| Fat | 5 | 21 | 11 | 16 |
| Carbohydrate | 52 | 10 | 6 | 48 |

**Example 7.1b**  A farm needs to purchase and combine four different types of food to obtain a mix for its pigs. It is required that at least 24% of the feed mix be protein, at least 12% should be fat, and exactly 32% should be carbohydrate. In addition, the percentage of fat should not exceed 0.6 times the percentage of protein. Table 7.1 gives the percentages of nutrient contents of the four different types of food.

If a total of five tons of food is needed, set up the problem of determining how much of each type should be purchased so that all constraints are satisfied at a minimal total cost, given that the cost (in dollars per ton) of the food types 1 to 4 are (respectively) 82, 144, 98, and 58.

*Solution.* If we let $x_i$ denote the number of tons of food item $i$ to be purchased, $i = 1, 2, 3, 4$, then the cost of the purchase is

$$82x_1 + 144x_2 + 98x_3 + 58x_4.$$

The amount of protein (in tons) in this mix would be

$$0.14x_1 + 0.52x_2 + 0.26x_3 + 0.06x_4,$$

which, to satisfy the protein requirement, would have to be at least $5(0.24) = 1.2$. As similar requirements must be met for the other nutrients, the mathematical problem would be to choose $x_1, x_2, x_3, x_4$, to

$$\text{minimize } 82x_1 + 144x_2 + 98x_3 + 58x_4$$

subject to the following constraints:

$$0.14x_1 + 0.52x_2 + 0.26x_3 + 0.06x_4 \geq 1.2,$$
$$0.05x_1 + 0.21x_2 + 0.11x_3 + 0.16x_4 \geq 0.6,$$
$$0.52x_1 + 0.10x_2 + 0.06x_3 + 0.48x_4 = 1.6,$$

$$0.05x_1 + 0.21x_2 + 0.11x_3 + 0.16x_4$$

$$\leq 0.6[0.14x_1 + 0.52x_2 + 0.26x_3 + 0.06x_4];$$

$$x_1 + x_2 + x_3 + x_4 = 5. \qquad \square$$

Any expression of the form

$$c_0 + c_1x_1 + \cdots + c_rx_r,$$

where $c_0, \ldots, c_r$ are specified constants, is said to be a *linear* function of the variables $x_1, \ldots, x_r$. For instance, whereas

$$x_1 + 3x_2 + 3x_3$$

is a linear function of the variables $x_1, x_2, x_3$, the expressions

$$x_1^2 + 2x_2 + x_3 \quad \text{and} \quad x_1x_2 + x_3$$

are not (the former because of the term $x_1^2$ and the latter because of the term $x_1x_2$). Thus, Examples 7.1a and 7.1b are both concerned with optimizing (maximizing or minimizing) linear functions of certain variables subject to linear constraints on these variables. Such optimization problems are known as *linear programs*.

The following optimization problem is called the *standard linear programming problem*. For given constants $c_i$ ($i = 1, \ldots, n$), $b_j$ ($j = 1, \ldots, m$), and $a_{i,j}$ ($i = 1, \ldots, n$, $j = 1, \ldots, m$), choose $x_1, x_2, \ldots, x_n$ to

$$\text{maximize} \sum_{i=1}^{n} c_i x_i$$

*subject to*

$$\sum_{i=1}^{n} a_{i,j} x_i \leq b_j, \quad j = 1, 2, \ldots, m,$$

$$x_i \geq 0, \quad i = 1, 2, \ldots, n.$$

A useful interpretation of the standard linear programming problem may be obtained by imagining a company that produces $n$ distinct items, each of which requires a certain amount of each of $m$ types of resources. Specifically, suppose that each unit of item $i$ requires $a_{i,j}$ units of resource $j$ ($j = 1, \ldots, m$) for its production. Assuming that the company earns a profit of $c_i$ per unit production of item $i$ ($i = 1, \ldots, n$), its problem is to decide how much of each item to produce so as to maximize its total profit, subject to the constraint that it has a total of $b_j$ units of resource $j$ available for each $j = 1, \ldots, m$. If we let $x_i$ ($i = 1, \ldots, n$) denote the number of units of item $i$ that the company produces, then $\sum_{i=1}^{n} c_i x_i$ represents the company's profit. Also, since producing $x_i$ units of item $i$ requires $x_i a_{i,j}$ units of resource $j$, it follows that $\sum_{i=1}^{n} a_{i,j} x_i$ represents the total amount of resource $j$ that will be used in producing these items. Thus, the company's problem is choose nonnegative numbers $x_1, \ldots, x_n$ to maximize its profit $\sum_{i=1}^{n} c_i x_i$, subject to the constraint that, for each resource $j$ ($j = 1, \ldots, m$), the amount of resource $j$ used up by this production program is less than or equal to $b_j$, the available amount of that resource. In other words, the company's problem is mathematically expressed by the standard linear programming problem.

Whereas the preceding interpretation is quite clear when all of the constants $a_{i,j}, b_i, c_j$ are nonnegative, it is interesting to see what the appropriate interpretation is when some of them are negative. To begin, suppose that $a_{i,j} < 0$ for some $i, j$. In this case, each unit of item $i$ produced requires a negative amount ($a_{i,j}$) of units of resource $j$; equivalently, each unit of item $i$ produced results in an additional $-a_{i,j}$ units of resource $j$. That is, $-a_{i,j}$ units of resource $j$ are obtained as a byproduct of the production of a unit of item $i$. If $b_j < 0$ then the company is required to end up with at least $-b_j$ units of resource $j$. That is, it must choose its production scheme so that at least this number of net units of resource $j$ is obtained. Also, if $c_i < 0$ then the company loses $-c_i$ per unit of item $i$ produced (although it still may be worthwhile to produce this type of item for the resources that are obtained from its production).

If there exist variables $x_1, \ldots, x_n$ that satisfy the constraints, then we say that the standard linear programming problem is *feasible*; if not, we say that it is *infeasible*. Software packages that solve the standard linear programming problem – by finding a solution when it exists, or by indicating that the problem is infeasible or that it is feasible but that no solution exists – are widely available, and we may hereafter suppose that any problem posed in the standard form can be explicitly solved.

## 7.2    Transforming to the Standard Form

Many problems concerned with optimizing a linear function of variables subject to linear constraints are not initially in the form of the standard linear programming problem and so must be transformed to that form. We now indicate how this is accomplished in a variety of cases.

### 7.2.1    *Minimization and Wrong-Way Inequality Constraints*

Suppose that we want to choose nonnegative values $x_1, \ldots, x_n$ to

$$\text{minimize } \sum_{i=1}^{n} c_i x_i$$

subject to a set of standard linear constraints. Because

$$\min \sum_{i=1}^{n} c_i x_i = -\max \sum_{i=1}^{n} -c_i x_i,$$

the problem can be solved by considering the standard linear program

$$\max \sum_{i=1}^{n} -c_i x_i,$$

subject to the linear constraints. A maximizing vector for this latter problem will also be a minimizing vector for the original problem.

A similar trick can be used if a constraint in the problem is given as

$$\sum_{i=1}^{n} a_{i,j} x_i \geq b_j.$$

Namely, multiplication by $-1$ yields the equivalent inequality

$$\sum_{i=1}^{n} -a_{i,j} x_i \leq -b_j,$$

which is in standard form.

A problem with an equality constraint of the form

$$\sum_{i=1}^{n} a_{i,j} x_i = b_j$$

can be transformed into standard form by writing the equality constraint as the pair of inequalities

$$\sum_{i=1}^{n} a_{i,j} x_i \le b_j,$$

$$\sum_{i=1}^{n} -a_{i,j} x_i \le -b_j.$$

## 7.2.2     *Problems with Variables Unconstrained in Sign*

Consider a linear programming problem in which some of the $x_i$ are not required to be nonnegative. Since any number can be treated as the difference between two nonnegative numbers, we can put such a problem into the standard form by replacing each unconstrained variable by two nonnegative variables whose difference is equal to the unconstrained variable. For instance, if the problem is in standard form except that $x_1$ is unconstrained in sign, then we can introduce a nonnegative variable $x_{n+1}$ and replace $x_1$ by $x_1 - x_{n+1}$. The problem becomes one of choosing $x_1, \ldots, x_n, x_{n+1}$ to

$$\text{maximize} \sum_{i=1}^{n+1} c_i x_i$$

*subject to*

$$\sum_{i=1}^{n+1} a_{i,j} x_i \le b_j, \quad j = 1, 2, \ldots, m,$$

$$x_i \ge 0, \quad i = 1, 2, \ldots, n+1,$$

where

$$c_{n+1} = -c_1 \quad \text{and} \quad a_{n+1,j} = -a_{1,j}.$$

If $x_i^o$ ($i = 1, \ldots, n+1$) is an optimal vector for this problem, then the optimal value of $x_1$ in the unconstrained version of the problem is $x_1 = x_1^o - x_{n+1}^o$.

**Example 7.2a**   For a given set of data pairs $w_i$, $y_i$ ($i = 1, \ldots, n$), suppose we want to find the straight line

$$y = a + bw$$

that best fits the data – in the sense that it minimizes the sum of the absolute differences between the values $y_i$ and the corresponding straight-line values $a + bw_i$. That is, we want to choose $a$ and $b$ to

$$\text{minimize } \sum_{i=1}^{n} |y_i - a - bw_i|.$$

If we let $x_i = y_i - a - bw_i$, then the problem becomes one of choosing $a, b, x_1, \ldots, x_n$ to

$$\text{minimize } \sum_{i=1}^{n} |x_i|.$$

Since the $x_i$ are unconstrained in sign, let us express them as the difference between two nonnegative variables $x_i'$ and $x_i''$; that is,

$$x_i = x_i' - x_i'',$$

where $x_i' \geq 0$ and $x_i'' \geq 0$. Now,

$$|x_i| \leq x_i' + x_i'',$$

with equality if at most one of the values $x_i'$ and $x_i''$ is nonzero. That is, whereas there are many nonnegative number pairs $x_i'$ and $x_i''$ such that $x_i = x_i' - x_i''$, the pair that minimizes $x_i' + x_i''$ is obtained by setting

$$x_i' = x_i^+ = \begin{cases} x_i & \text{if } x_i \geq 0, \\ 0 & \text{if } x_i \leq 0, \end{cases}$$

$$x_i'' = x_i^- = \begin{cases} -x_i & \text{if } x_i \leq 0, \\ 0 & \text{if } x_i \geq 0, \end{cases}$$

and, for this choice,

$$x_i' + x_i'' = |x_i|.$$

Thus, the problem of finding the values of $a$ and $b$ to minimize the sum of the absolute deviations from the straight line is one of choosing $a, b, x_i', x_i''$ ($i = 1, \ldots, n$) to

$$\text{minimize } \sum_{i=1}^{n} (x_i' + x_i'')$$

*subject to*

$$x_i' - x_i'' = y_i - a - bw_i, \quad i = 1, 2, \ldots, n,$$

$$x_i' \geq 0, \ x_i'' \geq 0, \quad i = 1, 2, \ldots, n.$$

The preceding linear program can then be put in standard form by

(a) transforming it from a minimization into a maximization problem,
(b) writing each equality constraint as two inequality constraints,
(c) introducing nonnegative variables $a, a', b, b'$ and replacing $a$ and $b$ by $a - a'$ and $b - b'$, respectively.   □

## 7.3    The Dual Linear Programming Problem

Consider the standard linear programming problem by using the interpretation of a company wanting to produce amounts of $n$ items from a stockpile of $m$ types of resources. Suppose now that a trader wishes to purchase all of the resources owned by the company. Let $y_j$ be the price per unit of resource $j$ that is offered by the trader, $j = 1, \ldots, m$. Since the production of a unit of item $i$ requires $a_{i,j}$ units of resource $j$ for each $j = 1, \ldots, m$, it follows that the trader is offering the amount $\sum_{j=1}^{m} a_{i,j} y_j$ for the package of resources needed to produce a unit of item $i$. Hence, if the trader chooses the prices so that, for each $i$, this amount exceeds the profit that the company makes from producing a unit of item $i$, then the company would be better off selling its resources. Hence, as the company possesses $b_j$ units of product $j$ ($j = 1, \ldots, m$), the trader will want to choose the prices to

$$\text{minimize } \sum_{j=1}^{m} b_j y_j$$

$$\text{subject to}$$

$$\sum_{j=1}^{m} a_{i,j} y_j \geq c_i, \quad i = 1, \ldots, n,$$

$$y_j \geq 0, \quad j = 1, \ldots, m.$$

The preceding linear programming problem is called the *dual* of the standard linear programming problem. It is usual to call the initial linear program whose dual is of interest the *primal* linear program.

**Example 7.3a**    Determine the dual of a primal linear program whose variables are all unconstrained in sign. That is, find the dual of

$$\text{maximize } \sum_{i=1}^{n} c_i x_i$$

*subject to*

$$\sum_{i=1}^{n} a_{i,j} x_i \le b_j, \quad j = 1, 2, \ldots, m.$$

**Solution.** Writing each $x_i$ as the difference of the nonnegative variables $x_i - x_{n+i}$ $(i = 1, \ldots, n)$ gives the equivalent linear program:

$$\text{maximize } \sum_{i=1}^{n} c_i (x_i - x_{n+i})$$

*subject to*

$$\sum_{i=1}^{n} a_{i,j} (x_i - x_{n+i}) \le b_j, \quad j = 1, 2, \ldots, m,$$

$$x_i \ge 0, \quad i = 1, 2, \ldots, 2n.$$

If we now let

$$c_{n+i} = -c_i \quad \text{and} \quad a_{n+i,j} = -a_{i,j} \quad (i = 1, \ldots, n),$$

then we can rewrite the preceding in the standard form:

$$\text{maximize } \sum_{i=1}^{2n} c_i x_i$$

*subject to*

$$\sum_{i=1}^{2n} a_{i,j} x_i \le b_j, \quad j = 1, 2, \ldots, m,$$

$$x_i \ge 0, \quad i = 1, 2, \ldots, 2n.$$

The dual linear program is thus to

$$\text{minimize } \sum_{j=1}^{m} b_j y_j$$

*subject to*

$$\sum_{j=1}^{m} a_{i,j} y_j \ge c_i, \quad i = 1, \ldots, 2n,$$

$$y_j \ge 0, \quad j = 1, \ldots, m.$$

Since $a_{n+i,j} = -a_{i,j}$ and $c_{n+i} = -c_i$ for $i = 1, \ldots, n$, it follows that the preceding inequality constraints are equivalent to

$$\sum_{j=1}^{m} a_{i,j} y_j \geq c_i, \quad i = 1, \ldots, n$$

and

$$-\sum_{j=1}^{m} a_{i,j} y_j \geq -c_i, \quad i = 1, \ldots, n.$$

Therefore, the dual program is to

$$\text{minimize } \sum_{j=1}^{m} b_j y_j$$

$$\text{subject to}$$

$$\sum_{j=1}^{m} a_{i,j} y_j = c_i, \quad i = 1, \ldots, n,$$

$$y_j \geq 0, \quad j = 1, \ldots, m.$$

In other words, when the primal variables are unconstrained in sign, the dual linear program has equality rather than inequality constraints.  □

The key theoretical result of linear programming is the *duality theorem*, which we state without proof.

**Proposition 7.3.1** (Duality Theorem of Linear Programming)  *If a standard and its dual linear program are both feasible, then they both have optimal solutions and the maximal value of the standard is equal to the minimal value of the dual. If either problem is infeasible, then the other does not have an optimal solution.*

**Example 7.3b**   Suppose that a farmer has $a$ acres of land on which he can plant wheat and rye. Where wheat is planted the farmer will attain a profit of either $w_1$ per acre if the year is dry or $w_2$ per acre if the year is wet. Similarly, where rye is planted the farmer will attain a profit of either $r_1$ per acre if the year is dry or $r_2$ per acre if the year is wet. How many acres should be used for wheat and for rye if the farmer wants to guarantee the largest possible profit in the coming year?

***Solution.*** Let $x_1$ and $x_2$ denote the number of acres devoted to wheat and rye, respectively. Since the farmer's profit will be $w_1 x_1 + r_1 x_2$ if the year is dry and $w_2 x_1 + r_2 x_2$ if the year is wet, it follows that the largest possible value of guaranteed profit – call it $x_3$ – is obtained by choosing $x_1, x_2, x_3$ to

$$\text{maximize } x_3$$

$$\textit{subject to}$$

$$w_1 x_1 + r_1 x_2 \geq x_3,$$

$$w_2 x_1 + r_2 x_2 \geq x_3,$$

$$x_1 + x_2 \leq a;$$

$$x_i \geq 0, \quad i = 1, 2, 3.$$

The dual of this problem (the verification is left as an exercise) is to choose $y_1, y_2, y_3$ to

$$\text{minimize } a y_3$$

$$\textit{subject to}$$

$$w_1 y_1 + w_2 y_2 \leq y_3,$$

$$r_1 y_1 + r_2 y_2 \leq y_3,$$

$$y_1 + y_2 \geq 1;$$

$$y_i \geq 0, \quad i = 1, 2, 3.$$

Since both linear programs are easily seen to be feasible, it follows from the duality theorem that the maximal value of the primal is equal to the minimal value of the dual.  □

A consequence of the duality theorem is the arbitrage theorem noted in Section 4.4.1. Recall that it refers to a situation in which there are $n$ wagers whose payoffs are determined by the outcome of an experiment whose possible outcomes are $1, 2, \ldots, m$. Specifically, if you bet wager $i$ at level $x$, then you win the amount $x r_i(j)$ if the outcome of the experiment is $j$. A betting strategy is a vector $\mathbf{x} = (x_1, \ldots, x_n)$, where each $x_i$ can be positive or negative (or zero) and with the interpretation that you simultaneously bet wager $i$ at level $x_i$ for each $i = 1, \ldots, n$. If the outcome of the experiment is $j$, then your winnings from the betting strategy **x** are equal to

$$\sum_{i=1}^{n} x_i r_i(j).$$

**Proposition 7.3.2** (Arbitrage Theorem)    *Exactly one of the following is true: Either*

(i) *there exists a probability vector* $\mathbf{p} = (p_1, \ldots, p_m)$ *for which*

$$\sum_{j=1}^{m} p_j r_i(j) = 0 \quad \text{for all } i = 1, \ldots, n$$

*or*

(ii) *there exists a betting strategy* $\mathbf{x} = (x_1, \ldots, x_n)$ *such that*

$$\sum_{i=1}^{n} x_i r_i(j) > 0 \quad \text{for all } j = 1, \ldots, m.$$

*That is, either there exists a probability vector under which all wagers have expected gain equal to 0, or else there is a betting strategy that always results in a positive win.*

***Proof.*** Let $x_{n+1}$ denote an amount that the gambler can be sure of winning, and consider the problem of maximizing this amount. If the gambler uses the betting strategy $(x_1, \ldots, x_n)$ then she will win $\sum_{i=1}^{n} x_i r_i(j)$ if the outcome of the experiment is $j$. Hence, she will want to choose her betting strategy $(x_1, \ldots, x_n)$ and $x_{n+1}$ so as to

$$\text{maximize } x_{n+1}$$

$$\text{subject to}$$

$$\sum_{i=1}^{n} x_i r_i(j) \geq x_{n+1}, \quad j = 1, \ldots, m.$$

Letting

$$a_{i,j} = -r_i(j), \quad i = 1, \ldots, n, \qquad a_{n+1,j} = 1,$$

we can rewrite the preceding as follows:

$$\text{maximize } x_{n+1}$$

$$\textit{subject to}$$

$$\sum_{i=1}^{n+1} a_{i,j} x_i \le 0, \quad j = 1, \ldots, m.$$

Note that the preceding linear program has $c_1 = c_2 = \cdots = c_n = 0$ and $c_{n+1} = 1$, upper-bound constraint values all equal to zero (i.e., all $b_j = 0$), and unconstrained variables $x_1, \ldots, x_{n+1}$. Hence, using the results of Example 7.3a (which shows that if the primal variables are unconstrained then the dual constraints are equality constraints), it follows that the dual of the preceding primal program is to choose variables $y_1, \ldots, y_m$ so as to

$$\text{minimize } 0$$

$$\textit{subject to}$$

$$\sum_{j=1}^{m} a_{i,j} y_j = 0, \quad i = 1, \ldots, n,$$

$$\sum_{j=1}^{m} a_{n+1,j} y_j = 1,$$

$$y_j \ge 0, \quad j = 1, \ldots, m.$$

Using the definitions of the quantities $a_{i,j}$ gives that this dual linear program is to

$$\text{minimize } 0$$

$$\textit{subject to}$$

$$\sum_{j=1}^{m} r_i(j) y_j = 0, \quad i = 1, \ldots, n,$$

$$\sum_{j=1}^{m} y_j = 1,$$

$$y_j \ge 0, \quad j = 1, \ldots, m.$$

Observe that this dual will be feasible and its minimal value will be 0 if and only if there exists a probability vector $(y_1, \ldots, y_m)$ under which all wagers have expected return 0. The primal problem is feasible because $x_i = 0$ $(i = 1, \ldots, n + 1)$ satisfies its constraints, so it follows from the duality theorem that if the dual problem is also feasible then the optimal value of the primal is 0 and thus no sure win is possible. On the other hand, if the dual is infeasible then it follows from the duality theorem that there is no optimal solution of the primal. But this implies that 0 is not the optimal solution, and thus there is a betting scheme whose minimal return is positive. (The reason there is no primal optimal solution when the dual is infeasible is because the primal is unbounded in this case. That is, if there is a betting scheme $\mathbf{x}$ that gives a guaranteed return of at least $v > 0$, then $c\mathbf{x}$ yields a guaranteed return of at least $cv$.)                                                                □

## 7.4    Game Theory

Consider the following game played by players I and II. They are presented with a matrix of $nm$ values $r_{i,j}$, where $i$ ranges from 1 to $n$ and $j$ from 1 to $m$. Player I then chooses one of the values $1, \ldots, n$ while player II simultaneously chooses one of the values $1, \ldots, m$. If player I chooses $i$ and player II chooses $j$, then player I receives the amount $r_{i,j}$ from player II. Such a game is called a two-person *zero-sum* game since whatever is won by one of the players is lost by the other. Let us consider a few examples.

**Example 7.4a**    Suppose $n = m = 3$, with the following payoff matrix.

$$
\begin{array}{c}
\text{II} \\
\hline
\begin{array}{c}
\text{I}
\end{array}
\left|
\begin{array}{rrr}
7 & -4 & 1 \\
3 & 5 & 2 \\
0 & 10 & -1
\end{array}
\right.
\end{array}
$$

In this example, each player has three possible strategies. If player I plays her strategy 1 and player II chooses his strategy 1 then player I wins from player II the amount 7; if I chooses strategy 1 and II chooses strategy 2 then player I wins $-4$ from II (i.e., II wins 4 from I in this case); and so on.                                                                □

**Example 7.4b**    For our next example, suppose that each player has two possible strategies, with the payoff matrix as follows.

$$
\begin{array}{c}
 & \text{II} \\
\text{I} & \begin{array}{|cc}
5 & 3 \\
2 & 8
\end{array}
\end{array}
$$

Since all the payoffs are positive, this game is clearly favorable to player I.    □

How much are the games in the preceding examples worth to player I? First, consider the game described in Example 7.4a and note that, by choosing strategy 2, player I can guarantee that she will win at least 2. Clearly the game is worth at least that amount to player I. On the other hand, by playing his strategy 3, player II can guarantee that player I does not win more than 2. Thus, it seems reasonable to suppose that the value of the game to player I is 2 and that strategy 1 is optimal for player I while strategy 3 is optimal for player II. However, the game in Example 7.4b is not so easily analyzed. By playing her strategy 1, player I can guarantee a win of at least 3 and so the value of the game to player I is at least 3. On the other hand, since player II can guarantee only that I will win no more than 5, it seems that the value of the game to player I should be somewhat larger than 3. Thus, in this case the value of the game to the two players appears uncertain.

The game of Example 7.4a was easy to analyze because its payoff matrix contained a *saddlepoint* – namely, a value that is simultaneously the minimum of its row (thus guaranteeing player I a win of at least that amount by choosing that row) and the maximum of its column (thus guaranteeing player II a loss of at most that amount by playing that column). The game of Example 7.4b does not have a saddlepoint, so its value is unclear.

The key to determining the value of a game is to allow the players to employ *randomized* (also known as *mixed*) strategies. That is, suppose we let the strategy of player I be a probability vector $\mathbf{p} = (p_1, p_2, \ldots, p_n)$, with the interpretation that player I will play her pure strategy $l$ with probability $p_i$, $i = 1, \ldots, n$. If player I uses this strategy and if player II plays his pure strategy $j$, then player I will win

$$r_{1,j} \text{ with probability } p_1,$$

$$r_{2,j} \text{ with probability } p_2,$$

$$\vdots$$

$$r_{i,j} \text{ with probability } p_i,$$

$$\vdots$$

$$r_{n,j} \text{ with probability } p_n.$$

Therefore, if player I uses the mixed strategy $\mathbf{p}$ then her *expected* winnings if player II chooses his pure strategy $j$ is $\sum_{i=1}^{n} p_i r_{i,j}$. Hence, by choosing the mixed strategy $\mathbf{p}$, player I can guarantee that she will have an expected win of at least

$$\min_{1 \le j \le m} \sum_{i=1}^{n} p_i r_{i,j}.$$

It thus follows that player I can guarantee herself an expected win of at least $v_1$, where

$$v_1 = \max_{\mathbf{p}} \min_{1 \le j \le m} \sum_{i=1}^{n} p_i r_{i,j}.$$

Similarly, if player II uses the mixed strategy given by the probability vector $\mathbf{y} = (y_1, \ldots, y_m)$ and if player I plays her strategy $i$, then player II's expected loss is $\sum_{j=1}^{m} y_j r_{i,j}$. Hence, the maximum expected loss of player II if he uses strategy $\mathbf{y}$ is

$$\max_{i=1,\ldots,n} \sum_{j=1}^{m} y_j r_{i,j}.$$

Therefore, player II has a strategy that guarantees that the expected win of player I is at most $v_2$, where

$$v_2 = \min_{\mathbf{y}} \max_{i=1,\ldots,n} \sum_{j=1}^{m} y_j r_{i,j}.$$

We now prove that $v_1$, the maximum of the minimal possible expected gain of player I, and $v_2$, the minimum of the maximal possible expected loss of player II, are equal. This is known as the *minimax* theorem of game theory.

**Proposition 7.4.1** (Minimax Theorem of Game Theory)

$$v_1 = v_2.$$

**Proof.** Since adding a constant to all the payoff values $r_{i,j}$ will increase both $v_1$ and $v_2$ by this amount, let us assume without loss of generality that all payoffs are positive. Letting $x_1, \ldots, x_n$ represent a probability vector and letting $x_{n+1}$ denote the minimum possible expected return for player I if she uses this probability vector, the maximum value of this minimum can be obtained from the following linear program:

$$\text{maximize } x_{n+1}$$

$$\textit{subject to}$$

$$\sum_{i=1}^{n} x_i r_{i,j} \geq x_{n+1}, \quad j = 1, 2, \ldots, m,$$

$$\sum_{i=1}^{n} x_i \leq 1, \quad -\sum_{i=1}^{n} x_i \leq -1,$$

$$x_i \geq 0, \quad i = 1, \ldots, n+1.$$

To put the preceding linear program into standard form, define $a_{i,j}$ ($i = 1, \ldots, n+1$, $j = 1, \ldots, m+2$) by

$$a_{i,j} = -r_{i,j}, \quad i \leq n, \ j \leq m,$$

$$a_{n+1,j} = 1, \quad j \leq m,$$

$$a_{i,m+1} = 1, \quad i \leq n,$$

$$a_{i,m+2} = -1, \quad i \leq n,$$

$$a_{n+1,j} = 0, \quad j = m+1, m+2,$$

and let

$$b_j = 0, \quad j \leq m,$$

$$b_{m+1} = 1,$$

$$b_{m+2} = -1,$$

$$c_i = 0, \quad i \leq n,$$

$$c_{n+1} = 1.$$

With these definitions, we see that player I's problem of determining the mixed strategy that maximizes her minimal expected return is equivalent to a standard linear program with $n + 1$ variables and $m + 2$ constraints. The dual of this standard linear program is to choose $y_1, \ldots, y_{m+2}$ to

$$\text{minimize} \sum_{j=1}^{m+2} b_j y_j$$

*subject to*

$$\sum_{j=1}^{m+2} a_{i,j} y_j \geq c_i, \quad i \leq n + 1,$$

$$y_j \geq 0, \quad j \leq m + 2.$$

But substituting back the values of $a_{i,j}$, $b_j$, and $c_i$ gives that the dual program is to choose $y_1, \ldots, y_{m+2}$ to

$$\text{minimize } y_{m+1} - y_{m+2}$$

*subject to*

$$-\sum_{j=1}^{m} r_{i,j} y_j + y_{m+1} - y_{m+2} \geq 0, \quad i \leq n,$$

$$\sum_{j=1}^{m} y_j \geq 1,$$

$$y_j \geq 0, \quad j \leq m + 2.$$

Letting $w = y_{m+1} - y_{m+2}$ and noting that (since $r_{i,j} > 0$) it is not necessary to consider any vectors $y_1, \ldots, y_m$ whose sum strictly exceeds 1 (for this would only increase the value of $w$), the dual can be written as follows:

$$\text{minimize } w$$

$$\textit{subject to}$$

$$\sum_{j=1}^{m} r_{i,j} y_j \leq w, \quad i \leq n,$$

$$\sum_{j=1}^{m} y_j = 1,$$

$$y_j \geq 0, \quad j \leq m.$$

Thus, the dual problem is just the minimax problem that is faced by player II. (That is, it is the mathematical statement of the problem in which player II is trying to choose the mixed strategy that minimizes his maximal expected loss.) Since both the primal and dual problems are clearly feasible, the minimax theorem follows from the duality theorem of linear programming. $\square$

## 7.5 Exercises

**Exercise 7.1** A company produces three types of fertilizer – A, B, and C. The production requires two different types of raw materials, with the following table giving the amounts of the raw materials needed to produce a ton of each type of fertilizer.

| Raw Material | Fertilizer Type | | |
|---|---|---|---|
| | A | B | C |
| 1 | 3 | 2.5 | 2 |
| 2 | 4 | 2 | 1 |

For example, it requires 3 units of raw material 1 and 4 units of material 2 to produce a ton of type-A fertilizer. The selling prices (dollars per ton) of the fertilizers is 275 for type A, 210 for type B, and 175 for type C. Assuming that the company has 1,250 units of raw material 1 and 1,000 units of material 2, set up the problem of determining how much of each type of fertilizer should be produced to maximize the amount of money received by the company.

**Exercise 7.2**  A company must ship its product, stored in $m$ different warehouses, to $n$ different destinations. The supply at warehouse $i$ is $s_i$, and the demand at destination $j$ is $d_j$. The cost of shipping $x$ units of the product from warehouse $i$ to destination $j$ is $xc(i, j)$. Set up the problem of determining how many units to ship from each warehouse to each destination – so as to meet all demands at a minimal cost – as a linear program. Also, give the necessary and sufficient conditions for this linear program to be feasible.

**Exercise 7.3**  Explain how an $n$-variable, $m$-constraint linear programming problem in which one of the variables is unconstrained in sign can be solved by solving two linear programs, each with $n$ variables and $m$ constraints. Why might this method be, in general, inferior to the technique introduced in Section 7.2.2?

*Hint:* If all $n$ variables are unconstrained, how many linear programs with $n$ variables and $m$ constraints would have to be solved?

**Exercise 7.4**  The number of employees needed by the post office on the different days of the week are as follows.

| Day | Number Needed |
| --- | --- |
| Monday | 18 |
| Tuesday | 12 |
| Wednesday | 14 |
| Thursday | 20 |
| Friday | 14 |
| Saturday | 17 |
| Sunday | 10 |

Employee contracts require that an employee must work five consecutive days and then receive two days off. Formulate a linear program that will enable the post office to minimize the number of employees needed to meet its requirements.

**Exercise 7.5** At the end of each of the next four months, a customer requires (respectively) 75, 100, 120, and 80 units of a certain commodity. Production costs per unit of the commodity vary from month to month and will be 6, 9, 5, and 7 for the next four months. Assuming that the monthly storage cost per item is 2.5, formulate a linear program whose solution yields the number of units to be produced in each of the next four months so as to meet the customer's requirements at minimal cost.

**Exercise 7.6** A project entails doing tasks $T_1, T_2, T_3, T_4, T_5$. However, some of the tasks cannot be started before certain others have been completed. With $T_i < T_j$ meaning that task $T_i$ must be completed before $T_j$ can be undertaken, suppose that

$$T_1 < T_3, \quad T_2 < T_4, \quad T_1 < T_4, \quad T_3 < T_5, \quad T_4 < T_5.$$

If $C_j$ represents the time it takes to complete task $T_j$ once it is begun, formulate (as a linear program) the problem of finding the minimal time in which the project can be completed.

**Exercise 7.7** Find the dual of the linear program of Exercise 7.1.

**Exercise 7.8** Find the dual of the linear program of Exercise 7.2.

**Exercise 7.9** Show that the dual of the dual is the original linear program. That is, take a standard linear program, express its dual in standard form, and then show that the dual of this latter linear program is the original standard linear program.

**Exercise 7.10** Set up a linear program that can be used to find the best linear fit to the following set of data pairs.

| $x$ | $y$ |
|---|---|
| 5 | 20 |
| 2 | 26 |
| 10 | 14 |
| 8 | 13 |
| 3 | 28 |

**Exercise 7.11**  Verify that the correct dual problem is given in Example 7.3b.

**Exercise 7.12**  Consider a horse race involving $n$ horses, with $o_i$ being the quoted odds against horse $i$ winning the race. That is, if you bet $x$ on horse $i$ then you win $xo_i$ if horse $i$ wins or lose $x$ if horse $i$ loses; $x$ can be either positive, negative, or zero, and a negative win (loss) is a loss (win). A betting strategy is a vector $\mathbf{x} = (x_1, \ldots, x_n)$, with the interpretation that you simultaneously place bets $x_i$ on horses $i = 1, \ldots, n$. Assuming that you are not allowed to have a (partial) loss of more than 1 on any bet relating to a single horse (i.e., a positive $x_i$ cannot exceed 1 and the absolute value of a negative $x_i$ cannot exceed $1/o_i$), give a linear program that can be used to find the strategy that guarantees the largest possible win.

**Exercise 7.13**  The following is a payoff matrix for a game in which both players have three different choices.

$$
\text{I} \quad
\begin{array}{ccc}
& \text{II} & \\
\hline
3 & 2 & -5 \\
1 & 8 & -10 \\
5 & 3 & -4 \\
\end{array}
$$

Tell which row player I should never choose and explain why.

**Exercise 7.14**  The following is a game payoff matrix.

$$
\text{I} \quad
\begin{array}{ccc}
& \text{II} & \\
\hline
12 & 4 & 8 \\
23 & 2 & 0 \\
36 & 1 & -6 \\
\end{array}
$$

Explain why the optimal strategy for player I has $p_2 = 0$; that is, explain why player I should never choose row 2.

*Hint:* What "randomization" rule is better than choosing row 2?

# 8. Sorting and Searching

## 8.1  Introduction to Sorting

Suppose we are presented with distinct values $x_1, x_2, \ldots, x_n$ that we desire to put in increasing order or (as is commonly stated) to *sort*. Probably the simplest sorting algorithm is to first find the smallest of these values by comparing $x_1$ with $x_2$, then comparing the smaller of these two with $x_3$, then the smaller of those two with $x_4$, and so on. After discovering the smallest value in this manner, we then find the second smallest value by repeating the process on the remaining $n - 1$ values, and so on. This method is called *selection sort* because it works by repeatedly selecting the smallest remaining value. Generally speaking, the amount of time that it takes an algorithm to sort a set of $n$ values will be proportional to the number of comparisons that are made. As the selection sort always requires

$$(n - 1) + (n - 2) + \cdots + 1 = \frac{n(n - 1)}{2}$$

comparisons, it follows that, for $n$ large, approximately $n^2/2$ comparisons are required.

## 8.2  The Bubble Sort

The bubble sort is another sorting algorithm. Starting with any initial ordering, it sequentially passes through the elements of this ordering, interchanging any pair that it finds out of order. That is, the first and second values are compared and interchanged if the second is smaller; then the new value in second position is compared with the value in the third position and these values are interchanged if the former is larger than the latter; then the new value in the third position is compared with the value in the fourth position; and so on until a comparison is made with the final value in the sequence and an interchange (if necessary) is made. At this point the "first pass" through the list is said to have occurred. This process is then repeated for the new ordering, and this continues until the values are sorted. For instance, if the initial ordering of values is

$$5\ 3\ 8\ 7\ 0\ 9\ 6\ 4\ 1,$$

then (with the bar indicating the value that is to be compared with its immediate follower) the successive orderings in the first pass are as follows:

$$3\ \bar{5}\ 8\ 7\ 0\ 9\ 6\ 4\ 1,$$

$$3\ 5\ \bar{8}\ 7\ 0\ 9\ 6\ 4\ 1,$$

$$3\ 5\ 7\ \bar{8}\ 0\ 9\ 6\ 4\ 1,$$

$$3\ 5\ 7\ 0\ \bar{8}\ 9\ 6\ 4\ 1,$$

$$3\ 5\ 7\ 0\ 8\ \bar{9}\ 6\ 4\ 1,$$

$$3\ 5\ 7\ 0\ 8\ 6\ \bar{9}\ 4\ 1,$$

$$3\ 5\ 7\ 0\ 8\ 6\ 4\ \bar{9}\ 1,$$

$$3\ 5\ 7\ 0\ 8\ 9\ 4\ 1\ 9.$$

It is easy to see that, after the first pass, the largest value will be the final value. As a result, the second pass does not need to consider the final value of the sequence and so will always result in the two largest values being in their correct positions. Similarly, the third pass through the list need not consider the final two values and will necessarily end with the final three values being the three largest values in the correct order, and so on. This algorithm is known as the *bubble sort* because the way in which small values move up to the front of the list is reminiscent of the way bubbles rise in a liquid.

The bubble sort algorithm ends either when no interchanges occur in a pass or when a total of $n - 1$ passes have been made. The $i$th pass requires a total of $n - i$ comparisons, so it follows that bubble sort requires $n - 1 + n - 2 + \cdots + 1 = n(n - 1)/2$ comparisons in the worst case. If we let 1 stand for the smallest value, 2 for the second smallest, and so on, then this worst case will occur if the initial ordering is

$$n, n - 1, n - 2, \ldots, 3, 2, 1.$$

However, since there is no particular reason to believe that the initial ordering will have the elements in decreasing order of their values, it makes sense to consider the *average* number of comparisons needed when the initial ordering is random. So, supposing that the initial order is equally likely to be any of the $n!$ orderings, let $X$ denote the number of comparisons needed by bubble sort and consider $E[X]$, the expected value of $X$. Although it is difficult to explicitly compute $E[X]$, we will be able to obtain bounds. First, since $X \leq n(n-1)/2$ for every initial ordering, we have

$$E[X] \leq \frac{n(n-1)}{2}. \tag{8.1}$$

To obtain a bound in the other direction, we need the concept of the number of inversions of a permutation. For any permutation $i_1, i_2, \ldots, i_n$ of $1, 2, \ldots, n$, we say that the ordered pair $(i, j)$ is an *inversion* of the permutation if $i < j$ and $j$ precedes $i$ in the permutation. For instance, the permutation

$$2, 4, 1, 5, 6, 3$$

has five inversions: $(1, 2)$, $(1, 4)$, $(3, 4)$, $(3, 5)$, and $(3, 6)$. Since the values of each inversion pair will eventually have to be interchanged (and thus compared), it follows that the number of comparisons made by the bubble sort is at least as large as the number of inversions of the initial ordering. That is, if $I$ denotes this number of inversions, then

$$X \geq I,$$

which implies that

$$E[X] \geq E[I]. \tag{8.2}$$

But if, for $i < j$, we let

$$I(i, j) = \begin{cases} 1 & \text{if } (i, j) \text{ is an inversion of the initial ordering,} \\ 0 & \text{otherwise,} \end{cases}$$

then it follows that

$$I = \sum_{j} \sum_{i<j} I(i, j).$$

Hence, using the fact that the expected value of a sum of random variables is equal to the sum of the expectations, we see that

$$E[I] = \sum_{j} \sum_{i<j} E[I(i, j)]. \tag{8.3}$$

Now, for $i < j$,

$$E[I(i, j)] = P\{j \text{ precedes } i \text{ in the initial ordering}\}.$$

But if the initial ordering is equally likely to be any of the $n!$ orderings, then it is as equally likely that $i$ precedes $j$ as it is that $j$ precedes $i$, implying that

$$E[I(i, j)] = 1/2.$$

Hence, since there are $\binom{n}{2}$ pairs $i, j$ for which $i < j$, from equation (8.3) we have that

$$E[I] = \frac{\binom{n}{2}}{2} = \frac{n(n-1)}{4}.$$

From (8.1), (8.2), and the preceding, we obtain

$$\frac{n(n-1)}{4} \leq E[X] \leq \frac{n(n-1)}{2}.$$

Thus, for large $n$, the average number of comparisons needed by a bubble sort to sort $n$ values is roughly between $n^2/4$ and $n^2/2$. Therefore, although bubble sort is more efficient than selection sort, it still requires on the order of $n^2$ comparisons. In order to obtain additional improvement, we must consider a different type of sorting algorithm.

## 8.3     The Quicksort Algorithm

Suppose again that we desire to sort the values $x_1, x_2, \ldots, x_n$. The *quicksort* algorithm is as follows. When $n = 2$, the algorithm compares the two values and puts them in the appropriate order. When $n > 2$, one of the values is chosen, say it is $x_i$, and then all of the other values are compared with $x_i$. Those smaller than $x_i$ are put in a bracket to the left of $x_i$, and those larger than $x_i$ are put in a bracket to the right of $x_i$. The algorithm then repeats itself on these brackets, continuing until all values have been sorted.

For example, suppose that we desire to sort the following ten distinct values:

$$5 \ 9 \ 3 \ 10 \ 11 \ 14 \ 8 \ 4 \ 17 \ 6.$$

One of these values is now chosen, say it is 10. We then compare each of the other values to 10, putting those less than 10 in a bracket to the left of 10 and putting those greater than 10 in a bracket to the right of 10. This gives

[5 9 3 8 4 6] 10 [11 14 17].

We now focus on a bracketed group that contains more than a single value (say, the one to the left of the 10) and choose one of its values – say 6 is chosen. Comparing each of the values in this bracket with 6 and putting the smaller ones in a bracket to the left of 6 and the larger ones in a bracket to the right of 6 gives

[5 3 4] 6 [9 8] 10 [11 14 17].

If we now consider the leftmost bracket and choose (say) the value 4 for comparison, then the next iteration yields

[3] 4 [5] 6 [9 8] 10 [11 14 17].

This process continues until all bracketed groups contain a single value only.

It is intuitively clear that the worst case occurs when every comparison value chosen is an extreme value – either the smallest or largest in its bracket. In this worst-case scenario, it is easy to see that the number of comparisons needed is $n(n-1)/2$. However, one obtains a better indication of the usefulness of the quicksort algorithm by determining the average number of comparisons needed when the comparison values are randomly chosen. So, let us suppose that each comparison value chosen from a bracket is equally likely to be any of the values in that bracket. (This is equivalent to assuming that the initial ordering is random and that the comparison value is always taken to be the first value to have been put in the bracket.) Let $X$ denote the number of comparisons needed. To compute $E[X]$, we will first express $X$ as the sum of other random variables in the following manner. To begin, give the following names to the values that are to be sorted: let 1 denote the smallest, let 2 denote the second smallest, and so on. Then, for $1 \le i < j \le n$, let $I(i, j)$ equal 1 if $i$ and $j$ are ever directly compared, and let it equal 0 otherwise. Summing these variables over all $i < j$ gives the total number of comparisons:

$$X = \sum_{j=2}^{n} \sum_{i=1}^{j-1} I(i, j),$$

which implies that

$$E[X] = E\left[\sum_{j=2}^{n}\sum_{i=1}^{j-1} I(i,j)\right]$$

$$= \sum_{j=2}^{n}\sum_{i=1}^{j-1} E[I(i,j)]$$

$$= \sum_{j=2}^{n}\sum_{i=1}^{j-1} P\{i \text{ and } j \text{ are ever compared}\}.$$

To determine the probability that $i$ and $j$ are ever compared, note that the values $i, i+1, \ldots, j-1, j$ will initially be in the same bracket (since all values are initially in the same bracket) and will remain in the same bracket if the number chosen for the first comparison is not between $i$ and $j$. For instance, if the comparison number is larger than $j$, then all the values $i, i+1, \ldots, j-1, j$ will go in a bracket to the left of the comparison number, and if it is smaller than $i$ then they will all go in a bracket to the right. Thus all the values $i, i+1, \ldots, j-1, j$ will remain in the same bracket until the first time that one of them is chosen as a comparison value. At that point, all the other values between $i$ and $j$ will be compared with this comparison value. Now, if this comparison value is neither $i$ nor $j$ then, upon comparison with it, $i$ will go into a left bracket and $j$ into a right bracket; consequently $i$ and $j$ will never be compared. On the other hand, if the comparison value of the set $i, i+1, \ldots, j-1, j$ is either $i$ or $j$, then there will be a direct comparison between $i$ and $j$. Now, given that the comparison value is one of the values between $i$ and $j$, it follows that it is equally likely to be any of these $j-i+1$ values; thus the probability that it is either $i$ or $j$ is $2/(j-i+1)$. Therefore, we may conclude that

$$P\{i \text{ and } j \text{ are ever compared}\} = \frac{2}{j-i+1}.$$

Consequently, we see that

$$E[X] = \sum_{j=2}^{n}\sum_{i=1}^{j-1} \frac{2}{j-i+1}$$

$$= 2\sum_{j=2}^{n}\sum_{k=2}^{j} \frac{1}{k} \quad \text{(by letting } k = j-i+1\text{)}$$

$$= 2 \sum_{k=2}^{n} \sum_{j=k}^{n} \frac{1}{k} \quad \text{(by interchanging the order of summation)}$$

$$= 2 \sum_{k=2}^{n} \frac{n-k+1}{k}$$

$$= 2(n+1) \sum_{k=2}^{n} \frac{1}{k} - 2(n-1).$$

Using the approximation that, for large $n$,

$$\sum_{k=2}^{n} \frac{1}{k} \approx \log(n),$$

we see (upon ignoring the linear term $2(n-1)$) that the quicksort algorithm requires, on average, approximately $2n \log(n)$ comparisons to sort $n$ values. For large $n$, this is much less than what is needed under either selection sort or bubble sort.

## 8.4    Merge Sorts

Suppose we are presented with two sorted lists and are asked to merge them into a single sorted list. To accomplish this, compare the smallest elements of both lists, and remove the smaller of these two from its list. Continue to repeat this operation until one of the lists is empty. The elements in the order of their removal, with the elements remaining on the nonempty list put at the end, constitutes an ordering of all the elements. If the sorted lists are of sizes $k$ and $m$ then it follows that, in the worst case, it will take $k + m - 1$ comparisons to merge them into a single sorted list.

The preceding suggests another sorting strategy. Namely, if you want to sort $2m$ values, first divide them into two sets of size $m$, sort each set, and then merge them into a single sorted list. To sort each sublist we can again divide them into two groups, then sort, and then merge. Indeed, we may continue to subdivide the sorting problem until there are only two items in a list; if we let $M(n)$ denote the number of comparisons needed to sort $n$ elements by this *merge sort* method, then

$$M(2^k) = 2^k - 1 + 2M(2^{k-1}),$$

where the term $2^k - 1$ refers to the number of comparisons needed to merge the two lists of size $2^{k-1}$. Using the preceding equation, we obtain

$$M(2^k) = 2^k - 1 + 2M(2^{k-1})$$
$$= 2^k - 1 + 2[2^{k-1} - 1 + 2M(2^{k-2})]$$
$$= 2 \cdot 2^k - (1 + 2) + 2^2 M(2^{k-2})$$
$$= 2 \cdot 2^k - (1 + 2) + 2^2[2^{k-2} - 1 + 2M(2^{k-3})]$$
$$= 3 \cdot 2^k - (1 + 2 + 2^2) + 2^3 M(2^{k-3})$$
$$\vdots$$
$$= (k - 1) \cdot 2^k - (1 + \cdots + 2^{k-2}) + 2^{k-1} M(2)$$
$$= (k - 1) \cdot 2^k - (2^{k-1} - 1) + 2^{k-1}.$$

Therefore, we see that

$$M(2^k) = (k - 1)2^k + 1.$$

Now, if $n = 2^k$ then

$$k = \log_2(n).$$

Thus, for $n$ large,

$$M(n) \approx n \log_2(n).$$

Hence, the merge sort algorithm requires about $n \log_2(n)$ comparisons to sort $n$ values. Although this is comparable to the number needed by quicksort, the implementation of merge sort is more involved than that of quicksort.

## 8.5    Sequential Searching

Suppose that we must determine if an item is one of the types $1, 2, \ldots, n$ or instead is of a different type. In addition, suppose that determining if an item is type $i$ requires comparison with a standard type-$i$ item, with the result of the comparison being that we learn whether the item is or is not a type $i$. If the item is not of any of these $n$ types then we will say that it has type $n + 1$. A straightforward approach that can be used to determine the type of an item is to select a permutation $i_1, i_2, \ldots, i_n$ of

$1, 2, \ldots, n$ and then check in sequence whether it is type $i_1$, or $i_2$, ..., or $i_n$. This will require a total of $n$ checks in the worst case (where the item is not any of the $n$ types).

Whereas we cannot say that one permutation is better than another from the point of view of a worst-case comparison, this is not the case when we suppose that there are known probabilities concerning the type of the new item. For in this latter case, permutations can be compared according to the average number of comparisons needed to categorize the item when the item types are checked in the order of the permutation. It turns out that, if $p_i$ is the probability that a new item is of type $i$ ($i = 1, \ldots, n$, $\sum_i p_i \le 1$), then the best sequential ordering is to order the types in decreasing order of their probabilities. That is, one should first check whether the new item is of the type having largest probability; if not, then check if it is of the type having second largest probability, and so on. This intuitive result is now formally proven.

**Proposition 8.5.1** *Suppose that the types are renumbered so that*

$$p_1 \ge p_2 \ge \cdots \ge p_n.$$

*The expected number of comparisons needed to determine the type of a new item is minimized by using the permutation* $1, 2, \ldots, n$.

**Proof.** Let $q = 1 - \sum_{i=1}^{n} p_i$ be the probability that the new item is of type $n + 1$, and consider any permutation

$$\mathbf{P} = i_1, \ldots, i_k, j, i, i_{k+3}, \ldots, i_n,$$

where $i < j$. That is, for $i < j$, the permutation $\mathbf{P}$ calls for checking that an item is type $j$ immediately before it checks that it is of type $i$. Let us compare this with the permutation $\mathbf{P}'$ obtained from $\mathbf{P}$ by interchanging the positions of $i$ and $j$. That is,

$$\mathbf{P}' = i_1, \ldots, i_k, i, j, i_{k+3}, \ldots, i_n.$$

If we let $N$ denote the number of comparisons needed to determine the type of a new item, then its expected value when the permutation $\mathbf{P}$ is used is

$$E_{\mathbf{P}}[N] = \sum_{r=1}^{k} r p_{i_r} + (k+1)p_j + (k+2)p_i + \sum_{r=k+3}^{n} r p_{i_r} + nq,$$

whereas the corresponding expected value when $\mathbf{P}'$ is used is

$$E_{\mathbf{P}'}[N] = \sum_{r=1}^{k} rp_{i_r} + (k+1)p_i + (k+2)p_j + \sum_{r=k+3}^{n} rp_{i_r} + nq.$$

Hence,

$$\begin{aligned} E_{\mathbf{P}}[N] - E_{\mathbf{P}'}[N] &= (k+1)p_j + (k+2)p_i - (k+1)p_i - (k+2)p_j \\ &= p_i - p_j \\ &\geq 0. \end{aligned}$$

Therefore, any pairwise interchange in which the smaller numbered type is moved closer to the front will result in a new permutation whose expected number of comparisons is at least as small (and strictly smaller if the two types have different probabilities) as the original one. Now consider any permutation other than $1, 2, \ldots, n$. By a sequence of pairwise interchanges, each one resulting in a new permutation whose expected number of comparisons is at least as small as the one preceding it, we can obtain the permutation $1, 2, \ldots, n$. For instance, if $n = 3$ then each of the successive permutations is at least as good as the one preceding it:

$$(3, 1, 2), \quad (1, 3, 2), \quad (1, 2, 3).$$

Thus the expected number of comparisons needed when the permutation $1, 2, \ldots, n$ is used is at least as small as under any other permutation, which proves the result. □

## 8.6    Binary Searches and Rooted Trees

Suppose again that we have a collection of items of the types $1, 2, \ldots, n$ and, moreover, that there are distinct numbers $v_1, \ldots, v_n$ such that each type-$i$ item has number $v_i$ attached to it. Suppose further that, when a new item is compared with a type-$i$ item, we learn not only whether it is of this type but also, if it is not, whether its value is smaller or larger than $v_i$. Thus, by making use of this latter information we are able to eliminate from consideration either all types whose value is less than $v_i$ (when the value of the new item exceeds $v_i$) or all types whose value is greater than $v_i$ (when the value of the new item is less than $v_i$).

Under these assumptions, *binary search* is an efficient algorithm for determining whether a new item is one of the types $1, 2, \ldots, n$ or is a new type.

The binary search algorithm starts by sorting the values $v_1, \ldots, v_n$; say the sorted values are

$$v_{i_1}, v_{i_2}, \ldots, v_{i_j}, \ldots, v_{i_{n-1}}, v_{i_n}.$$

If the middle value in the sequence of sorted values is $v_{i_j}$ then the algorithm compares the new item with a type-$i_j$ item. If the new item has value $v_{i_j}$, we are finished; if it is greater, we next compare its value with the middle value of the sorted values

$$v_{i_{j+1}}, \ldots, v_{i_{n-1}}, v_{i_n}$$

and, if it is less, to the middle value of the sorted values

$$v_{i_1}, v_{i_2}, \ldots, v_{i_{j-1}},$$

and so on. For instance, if $n = 8$ and $v_i = i$, then we begin by comparing the value of the new item to 4 (or to 5, since there is no unique middle value in this case); if it is less than 4 we would next compare it with 2 and, if greater, with 6 (or 7), and so on.

Say that a new item that is not any of the types $1, \ldots, n$ is of type $n+1$. Let $B(n)$ denote the maximum number of comparisons needed by the binary search algorithm to determine the type of a new item when there are $n$ known types. Suppose first that $n$ is of the form $n = 2^k$. After the first comparison, if the item's type has not been determined then it will be known to be either new or one of at most $2^{k-1}$ specified known types. Therefore,

$$B(2^k) = 1 + B(2^{k-1}).$$

Starting with

$$B(1) = 1,$$

we see that

$$B(2) = 1 + 1 = 2,$$
$$B(2^2) = 1 + 2 = 3,$$
$$B(2^3) = 1 + 3 = 4,$$
$$B(2^4) = 1 + 4 = 5,$$

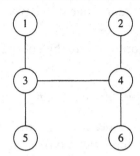

Figure 8.1: A Tree

and it is easily seen by induction that

$$B(2^k) = k + 1.$$

Thus, for $n$ large,

$$B(n) \approx \log_2(n).$$

Detailed instructions for implementing the binary search algorithm can be efficiently represented in a special type of directed graph known as a "rooted tree." Consider a tree (i.e., a connected graph without any cycles); specify one of its vertices, say $v_0$, and then give directions to the edges so that there are paths from the specified vertex to each of the other vertices of the tree. The resulting directed graph is called a *rooted tree,* and $v_0$ is called its *root.* For the original tree given in Figure 8.1, for example, two rooted versions (with roots 1 and 4, respectively) are shown in Figure 8.2.

When drawing a rooted tree, it is standard to place the root vertex at the top of the tree, to place vertices adjacent to the root vertex one level below the root, and so on. In addition, since it is understood that the edges point downward, it is common to suppress the edge direction arrows when drawing a rooted tree. Figure 8.3 gives the standard drawings of the rooted trees of Figure 8.2.

The root vertex of a rooted tree is said to be at level 0, those vertices on the next level are said to be at level 1, and so on. Thus the level of

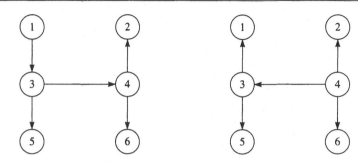

Figure 8.2: Two Rooted Trees

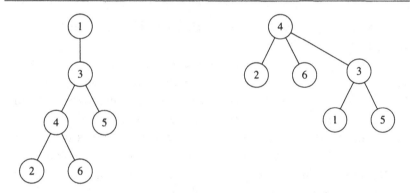

Figure 8.3: Standard Representations of Rooted Trees

a vertex is the length of the path from the root vertex to the vertex in question. For instance, in the rooted tree depicted on the left side of Figure 8.3, vertex 6 is at level 3; whereas for the tree on the right side of Figure 8.3, vertex 6 is at level 1. If $(i, j)$ is an edge of a rooted tree, we say that $j$ is a *child* of $i$ or that $i$ is a *parent* of $j$. Vertices that have children are called *internal* vertices, and those without are called *leaves*. For the rooted tree on the left side of Figure 8.3, vertices 2, 5, 6 are leaves and the others are internal vertices. If, for some $m$, each internal vertex of a rooted tree has $m$ children, we say that the rooted tree is an *m-ary*

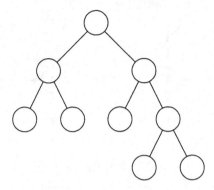

Figure 8.4: A Binary Tree

tree – when $m = 2$, a *binary* tree. The rooted tree of Figure 8.4 is a binary tree.

We can use a rooted tree to represent the instruction set of a binary search. The root of the tree is the first item type that is compared with the new item. If the value of the new item is less than the value of the comparison type, then the search moves to the left child of the comparison type, and if it has a larger value it moves to the right child. Figure 8.5 represents the binary search of a new item when there are 16 items and item type $j$ has the $j$th smallest value. It instructs one to first compare the new item with one of type 8; if its value is smaller than that of a type 8 then the next comparison is made with a type 4, and if it is larger then the next comparison is with a type 12, and so on.

An unsuccessful search will lead to one of the external square vertices, where the labels on these vertices give information about the unknown value $v$ of the item. The label $i^-$ means that $v \in (v_{i-1}, v_i)$ and the label $i^+$ that $v \in (v_i, v_{i+1})$, where $v_0 = -\infty$ and $v_{n+1} = \infty$.

The following example demonstrates another use of rooted trees.

**Example 8.6a**  A soccer team of 16 players and a coach use the telephone calling tree of Figure 8.6 to keep the players informed about upcoming games. The coach initiates the calls by calling three players, each of whom then calls another set of three players, and so on.

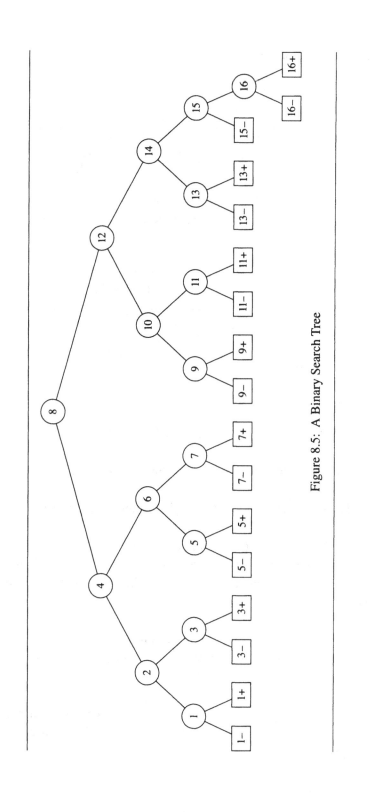

Figure 8.5: A Binary Search Tree

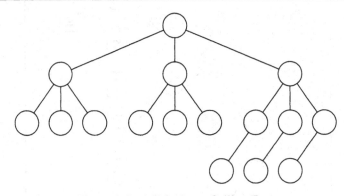

Figure 8.6:  A Telephone Calling Tree

## 8.7    Exercises

**Exercise 8.1**    Use selection sort to sort the values

3 12 4 22 17 9 55 62 38 23 27 18 5 30 19 74 64 15.

**Exercise 8.2**    Use bubble sort to sort the values in Exercise 8.1.

**Exercise 8.3**    Use quicksort to sort the values in Exercise 8.1.

**Exercise 8.4**    Suppose that the values to be sorted need not be distinct –
that is, some may be equal. Describe the quicksort algorithm in this situation. How many comparisons would it take quicksort to sort *n* equal
values?

**Exercise 8.5**    A sorting algorithm is said to be *stable* if the original
order is preserved when all *n* items have equal value. Is bubble sort stable? Is quicksort? Is merge sort?

**Exercise 8.6**    Suppose you are given a list of numbers that is in the correct order except for a few out-of-place values. Which sorting algorithm
would you use?

**Exercise 8.7**   Since quicksort works best when the comparison value chosen is the middle value, give a way of improving it by first randomly choosing three values before determining the comparison value.

**Exercise 8.8**   Devise an algorithm, in the spirit of quicksort, that can be used to find the $k$th smallest of a set of $n$ distinct numbers, $k \leq n$.

**Exercise 8.9**   How many vertices are there in an $m$-ary rooted tree with $i$ internal vertices?

**Exercise 8.10**   The *height* of a rooted tree is the length of the longest path of the tree (i.e., the maximal level of a vertex of the tree). Find an upper bound on the number of leaves of a $m$-ary tree of height $h$.

**Exercise 8.11**   Is the telephone calling tree of Example 8.6a a 3-ary tree? If not, how could it most easily be transformed into one?

# 9. Statistics

## 9.1 Introduction

It has become accepted that, in order to learn about something, you must first collect data. *Statistics* is the art of learning from data. It is concerned with the collection of data, its subsequent description, and its analysis, which often leads to the drawing of conclusions.

## 9.2 Frequency Tables and Graphs

A data set having a relatively small number of distinct values can be conveniently presented in a *frequency table*. For instance, Table 9.1 is a frequency table for a data set consisting of the starting yearly salaries (to the nearest thousand dollars) of 46 recently graduated students with a B.S. degree in computer science.

The frequency table tells us, among other things, that the lowest starting salary of $43,000 was received by four of the graduates, whereas the highest salary of $60,000 was received by a single student. The most common starting salary was $46,000, which was received by eight of the students.

Data from a frequency table can be graphically represented by plotting the distinct data values on the horizontal axis and indicating their frequencies by the heights of vertical segments. The graph is called a *line graph* if these segment are lines, or a *bar graph* if they are given added thickness. Figure 9.1 presents a bar graph for the data of Table 9.1.

When a data set has a large number of distinct values, we often divide these values into groupings, or *class intervals,* and then plot the number of data values falling in each class interval. A bar graph plot of class data, with the bars placed adjacent to each other, is called a *histogram.* Figure 9.2 presents a histogram of a data set consisting of the lifetimes of 200 incandescent lamps. It indicates, for example, that two of the

Table 9.1

| Starting Salary | Frequency |
|:---:|:---:|
| 43 | 4 |
| 44 | 3 |
| 45 | 5 |
| 46 | 8 |
| 47 | 6 |
| 48 | 4 |
| 50 | 6 |
| 52 | 5 |
| 55 | 4 |
| 60 | 1 |

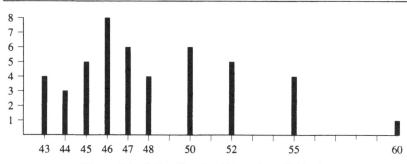

Figure 9.1: A Bar Graph of Data from Table 9.1

bulbs had lifetimes between 500 and 600 hours, five of the bulbs had lifetimes between 600 and 700 hours, twelve of the bulbs had lifetimes between 700 and 800 hours, and so on.

An efficient way of presenting a small to moderate-sized data set is to use a *stem-and-leaf plot*. Such a plot is obtained by first dividing each data value into two parts – its stem and its leaf. For instance, if the data are all two digit numbers, then we could let the stem part of a data value be its "tens" digit and let the leaf be its "ones" digit. Thus, for instance, the value 62 would be expressed as follows.

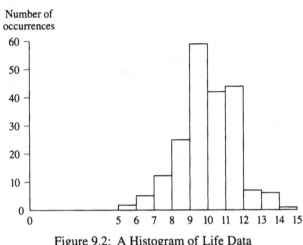

Figure 9.2: A Histogram of Life Data

| Stem | Leaf |
| --- | --- |
| 6 | 2 |

Likewise, the two data values 62 and 67 would be represented as follows.

| Stem | Leaf |
| --- | --- |
| 6 | 2, 7 |

**Example 9.2a**   The following stem-and-leaf plot gives the scores on a final examination in a course that used this text.

```
10 | 0
 9 | 1, 2, 6, 6
 8 | 0, 2, 3, 5, 5, 7, 7, 8, 9
 7 | 0, 0, 2, 4, 5, 5, 6, 7, 8, 8
 6 | 0, 5, 5, 8, 8
 5 | 0, 5, 5, 7
```

This stem-and-leaf plot tells us, among other things, that scores ranged from a low of 50 to a high of 100, and that there were ten scores in the 70s.

## 9.3     Summarizing Data Sets

In order to obtain a feel for large amounts of data, it is useful to be able to summarize the data by some suitably chosen measures. We now present some summarizing *statistics,* where a statistic is a numerical quantity whose value is determined by the data.

### 9.3.1     *Sample Mean, Sample Median, and Sample Mode*

Three statistics that are used for describing the center of a set of data values are the sample mean, the sample median, and the sample mode. The sample mean of the data set consisting of the $n$ numerical values $x_1, x_2, \ldots, x_n$ is the arithmetical average of these values.

**Definition**   The *sample mean,* designated by $\bar{x}$, is defined by

$$\bar{x} = \sum_{i=1}^{n} \frac{x_i}{n}.$$

The computation of the sample mean can often be simplified by noting that, for constants $a$ and $b$, if

$$y_i = ax_i + b, \quad i = 1, \ldots, n,$$

then the sample mean of the data set $y_1, \ldots, y_n$ is

$$\bar{y} = \sum_{i=1}^{n} \frac{ax_i + b}{n} = \sum_{i=1}^{n} \frac{ax_i}{n} + \sum_{i=1}^{n} \frac{b}{n} = a\bar{x} + b.$$

**Example 9.3a**   The winning scores in the U.S. Master's Golf Tournament in the years from 1982 to 1991 were as follows:

284, 280, 277, 282, 279, 285, 281, 283, 278, 277.

Find the sample mean of these scores.

*Solution.* Rather than directly adding these values, it is easier to first subtract 280 from each one to obtain the new values $y_i = x_i - 280$:

$$4, \; 0, \; -3, \; 2, \; -1, \; 5, \; 1, \; 3, \; -2, \; -3.$$

The arithmetic average of the transformed data set is

$$\bar{y} = 6/10,$$

so it follows that

$$\bar{x} = \bar{y} + 280 = 280.6. \qquad \qquad \Box$$

Sometimes we want to determine the sample mean of a data set that is presented in a frequency table listing the $k$ distinct values $v_1, \ldots, v_k$ having corresponding frequencies $f_1, \ldots, f_k$. Such a data set consists of $n = \sum_{i=1}^{k} f_i$ observations, with the value $v_i$ appearing $f_i$ times for each $i = 1, \ldots, k$. Therefore, the sample mean of these $n$ data values is

$$\bar{x} = \sum_{i=1}^{k} \frac{v_i f_i}{n}.$$

Rewriting this as

$$\bar{x} = \frac{f_1}{n} v_1 + \frac{f_2}{n} v_2 + \cdots + \frac{f_k}{n} v_k,$$

we see that the sample mean is a *weighted average* of the distinct values, where the weight given to the value $v_i$ is equal to the proportion of the $n$ data values that are equal to $v_i$, $i = 1, \ldots, k$.

Another statistic used to indicate the center of a data set is the sample median, defined as follows. Order the values of a data set of size $n$ from smallest to largest. If $n$ is odd, the *sample median* is the value in position $(n + 1)/2$; if $n$ is even, it is the average of the values in positions $n/2$ and $n/2 + 1$. Thus the sample median of a set of three values is the second smallest; of a set of four values, it is the average of the second and third smallest.

Another statistic that has been used to indicate the central tendency of a data set is the *sample mode*, defined to be the value that occurs with the greatest frequency. If no single value occurs most frequently, then all the values that occur at the highest frequency are called *modal values.*

**Example 9.3b** The following frequency table gives the values obtained in 40 rolls of a die.

| Value | Frequency |
|-------|-----------|
| 1 | 9 |
| 2 | 8 |
| 3 | 5 |
| 4 | 5 |
| 5 | 6 |
| 6 | 7 |

Find the sample mean, the sample median, and the sample mode.

*Solution.* The sample mean is

$$\bar{x} = (9 + 16 + 15 + 20 + 30 + 42)/40 = 3.05.$$

The sample median is the average of the 20th and 21st smallest values and is thus equal to 3. The sample mode is 1, the value that occurred most frequently. □

## 9.3.2 *Sample Variance and Sample Standard Deviation*

We have presented statistics that describe the central tendencies of a data set, yet we are also interested in ones that describe the "spread" or variability of the data values. A statistic that could be used for this purpose would be one that measures the average value of the squares of the distances between the data values and the sample mean. This is accomplished by the sample variance, which for technical reasons divides the sum of the squares of the differences by $n - 1$ rather than $n$, where $n$ is the size of the data set.

**Definition** The *sample variance* (call it $s^2$) of the data set $x_1, \ldots, x_n$ is defined by

$$s^2 = \sum_{i=1}^{n} \frac{(x_i - \bar{x})^2}{n - 1}.$$

The following algebraic identity is often useful for computing the sample variance:

$$\sum_{i=1}^{n}(x_i - \bar{x})^2 = \sum_{i=1}^{n}x_i^2 - n\bar{x}^2.$$

The identity is proven as follows:

$$\sum_{i=1}^{n}(x_i - \bar{x})^2 = \sum_{i=1}^{n}(x_i^2 - 2x_i\bar{x} + \bar{x}^2)$$

$$= \sum_{i=1}^{n}x_i^2 - 2\bar{x}\sum_{i=1}^{n}x_i + \sum_{i=1}^{n}\bar{x}^2$$

$$= \sum_{i=1}^{n}x_i^2 - 2n\bar{x}^2 + n\bar{x}^2$$

$$= \sum_{i=1}^{n}x_i^2 - n\bar{x}^2.$$

The computation of the sample variance can also be eased by noting that, if

$$y_i = a + bx_i, \quad i = 1, \ldots, n,$$

then $\bar{y} = a + b\bar{x}$ and so

$$\sum_{i=1}^{n}(y_i - \bar{y})^2 = b^2\sum_{i=1}^{n}(x_i - \bar{x})^2.$$

That is, if $s_y^2$ and $s_x^2$ are the respective sample variances, then

$$s_y^2 = b^2 s_x^2.$$

In other words, adding a constant to each data value does not change the sample variance; whereas multiplying each data value by a constant results in a new sample variance that is equal to the old one multiplied by the square of the constant.

**Example 9.3c**   The following data give the worldwide number of fatal airline accidents of commercially scheduled air transports in the years from 1985 to 1993.

| Year | Number of Accidents |
|------|------|
| 1985 | 22 |
| 1986 | 22 |
| 1987 | 26 |
| 1988 | 28 |
| 1989 | 27 |
| 1990 | 25 |
| 1991 | 30 |
| 1992 | 29 |
| 1993 | 24 |

Find the sample variance of the number of accidents in these years.

*Solution.* Let us start by subtracting 22 from each value to obtain the new data set

$$0, 0, 4, 6, 5, 3, 8, 7, 2.$$

Calling the transformed data $y_1, \ldots, y_9$, we have

$$\sum_{i=1}^{n} y_i = 35, \qquad \sum_{i=1}^{n} y_i^2 = 203.$$

Hence, since the sample variance of the transformed data is equal to that of the original data, we have (upon using the algebraic identity) that

$$s^2 = 203 - 9(35/9)^2 = 66.889. \qquad \square$$

The positive square root of the sample variance is called the *sample standard deviation*. That is, the sample standard deviation is

$$s = \sqrt{\sum_{i=1}^{n} \frac{(x_i - \bar{x})^2}{n - 1}}.$$

## 9.4    Chebyshev's Inequality

Let $\bar{x}$ and $s$ be the sample mean and sample standard deviation of a data set. Assuming that $s > 0$, Chebyshev's inequality implies that, for any value of $k \geq 1$, at least $100(1 - 1/k^2)$ percent of the data lie within the interval from $\bar{x} - ks$ to $\bar{x} + ks$. Letting $k = 3/2$, we obtain from Chebyshev's inequality that at least $100(5/9) = 55.56\%$ of the data from any data set lies within a distance $1.5s$ of the sample mean $\bar{x}$; letting $k = 2$ shows that at least 75% of the data lies within $2s$ of the sample mean; and letting $k = 3$ shows that at least $800/9 = 88.9\%$ of the data lies within three sample standard deviations of $\bar{x}$.

If the size of the data set is specified then Chebyshev's inequality can be sharpened, as shown in the following formal statement and proof.

**Theorem 9.4.1** (Chebyshev's Inequality)  *Let $\bar{x}$ and $s$ be the sample mean and sample standard deviation of the data set consisting of the data $x_1, \ldots, x_n$, where $s > 0$. Let*

$$S_k = \{i \ (1 \leq i \leq n) : |x_i - \bar{x}| < ks\},$$

*and let $|S_k|$ denote the number of elements in the set $S_k$. Then, for any $k \geq 1$,*

$$\frac{|S_k|}{n} \geq 1 - \frac{n-1}{nk^2} > 1 - \frac{1}{k^2}.$$

**Proof.**

$$(n-1)s^2 = \sum_{i=1}^{n}(x_i - \bar{x})^2$$

$$= \sum_{i \in S_k}(x_i - \bar{x})^2 + \sum_{i \notin S_k}(x_i - \bar{x})^2$$

$$\geq \sum_{i \notin S_k}(x_i - \bar{x})^2$$

$$\geq \sum_{i \notin S_k} k^2 s^2$$

$$= k^2 s^2 (n - |S_k|),$$

where the first inequality follows because all terms being summed are nonnegative; the second follows since $(x_i - \bar{x})^2 \geq k^2 s^2$ when $i \notin S_k$. Dividing both sides of the preceding inequality by $nk^2 s^2$ yields that

$$\frac{n-1}{nk^2} \geq 1 - \frac{|S_k|}{n},$$

and the result is proven.    □

Table 9.2

| Day | Temperature | Number of Defects |
|-----|-------------|-------------------|
| 1   | 24.2        | 25                |
| 2   | 22.7        | 31                |
| 3   | 30.5        | 36                |
| 4   | 28.6        | 33                |
| 5   | 25.5        | 19                |
| 6   | 32.0        | 24                |
| 7   | 28.6        | 27                |
| 8   | 26.5        | 25                |
| 9   | 25.3        | 16                |
| 10  | 26.0        | 14                |
| 11  | 24.4        | 22                |
| 12  | 24.8        | 23                |
| 13  | 20.6        | 20                |
| 14  | 25.1        | 25                |
| 15  | 21.4        | 25                |
| 16  | 23.7        | 23                |
| 17  | 23.9        | 27                |
| 18  | 25.2        | 30                |
| 19  | 27.4        | 33                |
| 20  | 28.3        | 32                |
| 21  | 28.8        | 35                |
| 22  | 26.6        | 24                |

## 9.5    Paired Data Sets and the Sample Correlation Coefficient

We are often concerned with data sets that consist of pairs of values that have some relationship to each other. If each element in such a data set has an $x$-value and a $y$-value, then we represent the $i$th data point by the pair $(x_i, y_i)$. For instance, in an attempt to determine the relationship between the daily midday temperature (measured in degrees Celsius) and the number of defective parts produced during that day, a company recorded the data presented in Table 9.2. For this data set, $x_i$ represents the temperature and $y_i$ the number of defective parts produced on day $i$.

A useful way of portraying a data set of paired values is to plot the data on a two-dimensional graph, with the $x$-axis representing the $x$-value of the data and the $y$-axis representing the $y$-value. Such a plot is called

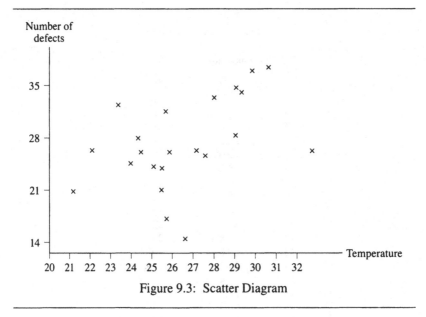

Figure 9.3:  Scatter Diagram

a *scatter diagram*. Figure 9.3 presents a scatter diagram for the data of
Table 9.2.

A question of interest concerning paired data sets is whether large
$x$-values tend to be paired with large $y$-values and small $x$-values with
small $y$-values; if this is not the case, we might then question whether
large values of one of the variables tend to be paired with small values
of the other. A rough answer to these questions can often be provided by
the scatter diagram. For instance, Figure 9.3 indicates that there appears
to be some connection between high temperatures and large numbers of
defective items. To obtain a quantitative measure of this relationship,
we will now develop a statistic that attempts to measure the degree to
which larger $x$-values go with larger $y$-values and smaller $x$-values with
smaller $y$-values.

Suppose that the data set consists of the paired values $(x_i, y_i)$, $i =$
$1, \ldots, n$. To obtain a statistic that can be used to measure the association
between the paired data values, let $\bar{x}$ and $\bar{y}$ denote the sample means of
the $x$-values and the $y$-values, respectively. Now, for data pair $i$, consider $x_i - \bar{x}$, the deviation of its $x$-value from the $x$ sample mean, and

$y_i - \bar{y}$, the deviation of its $y$-value from the $y$ sample mean. If $x_i$ is a large $x$-value, then it will be larger than the average value of all the $x$ and so the deviation $x_i - \bar{x}$ will be a positive value. Similarly, when $x_i$ is a small $x$-value, the deviation $x_i - \bar{x}$ will be a negative value. As the same is true for the deviations in $y$, we can conclude that when large (resp. small) values of the $x$-variable tend to be associated with large (resp. small) values of the $y$-variable the signs (either positive or negative) of $x_i - \bar{x}$ and $y_i - \bar{y}$ tend to be the same.

Now, if $x_i - \bar{x}$ and $y_i - \bar{y}$ have the same sign then their product $(x_i - \bar{x})(y_i - \bar{y})$ will be positive. It follows that, if large $x$-values tend to be paired with large $y$-values and small $x$-values with small $y$-values, then $\sum_{i=1}^{n}(x_i - \bar{x})(y_i - \bar{y})$ will tend to be a large positive number. In fact, not only will all the products have a positive sign when large (small) $x$-values are paired with large (small) $y$-values, but it also follows from Proposition 1.3.2 (Hardy's lemma) that the largest possible value of the sum of paired products will be obtained when the largest $x_i - \bar{x}$ is paired with the largest $y_i - \bar{y}$, the second largest $x_i - \bar{x}$ is paired with the second largest $y_i - \bar{y}$, and so on. Conversely, it follows that when large values of $x_i$ tend to be paired with *small* values of $y_i$, the signs of $x_i - \bar{x}$ and $y_i - \bar{y}$ will be opposite and so $\sum_{i=1}^{n}(x_i - \bar{x})(y_i - \bar{y})$ will be a large negative number.

In order to determine what it means for $\sum_{i=1}^{n}(x_i - \bar{x})(y_i - \bar{y})$ to be "large," we standardize this sum first by dividing by $n - 1$ and then by dividing by the product of the two sample standard deviations. The resulting statistic is called the *sample correlation coefficient*. That is, the sample coefficient $r$ of the data pairs $(x_i, y_i)$, $i = 1, \ldots, n$, is defined by

$$
r = \frac{\sum_{i=1}^{n}(x_i - \bar{x})(y_i - \bar{y})}{(n - 1)s_x s_y}
$$

$$
= \frac{\sum_{i=1}^{n}(x_i - \bar{x})(y_i - \bar{y})}{\sqrt{\sum_{i=1}^{n}(x_i - \bar{x})^2 \sum_{i=1}^{n}(y_i - \bar{y})^2}}.
$$

When $r > 0$ we say that the sample data pairs are *positively correlated*, and when $r < 0$ we say that they are *negatively correlated*.

We now list some properties of the sample correlation coefficient.

**Properties of r**

(1) $-1 \le r \le 1$.
(2) If for constants $a$ and $b$, with $b > 0$,

$$y_i = a + bx_i, \quad i = 1, \ldots, n,$$

then $r = 1$.
(3) If for constants $a$ and $b$, with $b < 0$,

$$y_i = a + bx_i, \quad i = 1, \ldots, n,$$

then $r = -1$.
(4) If $r$ is the sample correlation coefficient for the data pairs $(x_i, y_i)$, $i = 1, \ldots, n$, then it is also the sample correlation coefficient for the data pairs $(a + bx_i, c + dy_i)$, $i = 1, \ldots, n$, provided that the constants $b$ and $d$ have the same sign.

Property (1) is a direct consequence of the Cauchy–Schwarz inequality, which states that for any values $a_i, b_i$ $(i = 1, \ldots, n)$,

$$\left(\sum_{i=1}^{n} a_i b_i\right)^2 \le \sum_{i=1}^{n} a_i^2 \sum_{i=1}^{n} b_i^2.$$

(Simply let $a_i = x_i - \bar{x}$ and $b_i = y_i - \bar{y}$.) The verifications of the other properties of $r$ are straightforward and are left as exercises.

The absolute value of the sample correlation coefficient (i.e., $|r|$) is a measure of the strength of the linear relationship between the $x$ and $y$ values of a data pair. A value of $r$ equal to 1 indicates a perfect linear relation; a value around 0.8 indicates that a linear relation gives a relatively good fit to the data pairs; a value around 0.3 indicates that a linear relation gives a relatively poor fit to the data pairs. The sign of $r$ gives the direction of the relation: it is positive when the linear fit points upward and negative when it points downward. (The sample correlation coefficient for the data pairs of Table 9.2 is $r = 0.4189$.)

## 9.6    Testing Statistical Hypotheses

One often collects data so as to be able to draw some inferences about the process that generates the data. A common inference problem that is often considered is that of using the data to test a certain statistical hypothesis. For instance, suppose that a manufacturing process produces items that may or may not meet certain specifications. A common practice in such cases is to suppose that each item produced independently fails to meet the specifications with some unknown probability $p$. Suppose now that we are interested in testing the hypothesis that $p \leq p_0$ for some specified probability $p_0$. To test this hypothesis, we would randomly select a *sample* of (say, $n$) items and then determine how many of these do not meet the specifications. If $k$ of the $n$ did not meet the specifications, then we would want to reject the hypothesis that $p \leq p_0$ if obtaining as many as $k$ failures in $n$ trials would be very unlikely to occur when the failure probability is less than or equal to $p_0$. To determine exactly how unlikely such an event would be to occur, we would determine $P\{X \geq k\}$, where $X$ is the random number of failures that would occur in $n$ independent trials when each trial is a failure with probability $p_0$. In other words, $X$ is a binomial random variable with parameters $n$ and $p_0$ (see Example 3.5b), and $P\{X \geq k\}$ can be explicitly determined by using a standard program for determining probabilities for binomial random variables. If $P\{X \geq k\}$ is sufficiently small, then the hypothesis would be rejected. For example, if $p_0 = 0.1$, $n = 500$, and $k = 74$, then we would reject the hypothesis that the probability of a defective item is at most 0.1 if the probability is sufficiently small that as many as 74 defective items would have occurred in 500 trials when the probability of a defective item is 0.1. Now, for a binomial random variable $X$ with parameters 500 and 0.1,

$$P\{X \geq 74\} = 0.0005.$$

For such a small probability, the hypothesis is rejected.

## 9.7 Exercises

**Exercise 9.1** The following is a sample of prices, rounded to the nearest cent, charged per gallon of standard unleaded gasoline in the San Francisco Bay area in June 1997:

137, 139, 141, 137, 144, 141, 139, 137, 144, 141, 143, 143, 141.

Represent these data in

(a) a frequency table,
(b) a line graph.

**Exercise 9.2**   If a city government has a flat-rate income tax and is trying to estimate its total revenue from the tax, which statistic would it
be more interested in: the sample mean or the sample median? What if
it were thinking about constructing middle-income housing and wanted
to determine the proportion of its population able to afford it?

**Exercise 9.3**   The sample mean and sample variance of five data values are $\bar{x} = 104$ and $s^2 = 4$. If three of the data values are 102, 100, and
105, what are the other two values?

**Exercise 9.4**   An efficient way to compute the sample mean and sample variance of the data set $x_1, x_2, \ldots, x_n$ is as follows. Let

$$\bar{x}_j = \frac{\sum_{i=1}^{j} x_i}{j}, \quad j = 1, \ldots, n,$$

be the sample mean of the first $j$ data values, and let

$$s_j^2 = \frac{\sum_{i=1}^{j} (x_i - \bar{x}_j)^2}{j - 1}, \quad j = 2, \ldots, n,$$

be the sample variance of the first $j$ ($j \geq 2$) values. Then, with $s_1^2 = 0$,
it can be shown that

$$\bar{x}_{j+1} = \bar{x}_j + \frac{\bar{x}_{j+1} - \bar{x}_j}{j + 1}$$

and

$$s_{j+1}^2 = (1 - 1/j)s_j^2 + (j + 1)(\bar{x}_{j+1} - \bar{x}_j)^2.$$

(a) Use the preceding formulas to compute the sample mean and sample variance of the data values 3, 4, 7, 2, 9, 6.
(b) Verify your results in part (a) by computing as usual.
(c) Verify the given formula for $\bar{x}_{j+1}$ in terms of $\bar{x}_j$.
(d) Verify the given formula for $s_{j+1}^2$ in terms of $s_j^2$.

**Exercise 9.5**   The *sample 100p percentile* is that data value such that
at least 100$p$ percent of the data are less than or equal to it and at least

$100(1 - p)$ percent are greater than or equal to it. If two data values satisfy this condition, then the sample $100p$ percentile is the arithmetic average of these two values.

(a) If $np$ is not an integer, how would you determine the sample $100p$ percentile?
(b) What if $np$ is an integer?

**Exercise 9.6**  The sample 50th percentile is the sample median. Along with the sample 25th (called the *first quartile*) and the sample 75th percentile (called the *third quartile*), it makes up the sample quartiles. Find the sample quartiles of the data set of Table 9.1.

**Exercise 9.7**  The following are the grade-point averages of 30 students recently admitted to the graduate program in the department of Industrial Engineering and Operations Research at the University of California at Berkeley:

3.46, 3.72, 3.95, 3.55, 3.62, 3.80, 3.86, 3.71, 3.56, 3.49,

3.96, 3.90, 3.70, 3.61, 3.72, 3.65, 3.48, 3.87, 3.82, 3.91,

3.69, 3.67, 3.72, 3.66, 3.79, 3.75, 3.93, 3.74, 3.50, 3.83.

(a) Represent the preceding data in a stem-and-leaf plot.
(b) Calculate the sample mean $\bar{x}$.
(c) Calculate the sample standard deviation $s$.
(d) Determine the proportion of the data values that lie within $\bar{x} \pm 1.5s$ and compare this amount with the lower bound given by Chebyshev's inequality.
(e) Determine the proportion of the data values that lie within $\bar{x} \pm 2s$ and compare this amount with the lower bound given by Chebyshev's inequality.

**Exercise 9.8**  Verify properties (2) and (3) of the sample correlation coefficient.

**Exercise 9.9**  Verify property (4) of the sample correlation coefficient.

**Exercise 9.10**  If $r$ is the sample correlation coefficient of the data set $(x_i, y_i)$, $i = 1, \ldots, n$, what is the sample correlation coefficient of the data set $(x_i, -y_i)$, $i = 1, \ldots, n$?

**Exercise 9.11**  The daily temperatures specified in Table 9.2 are given in degrees Celsius. Given that the sample correlation coefficient for the data pairs of this table is $r = 0.4189$, what would it have been had the temperature been given in degrees Fahrenheit?

# 10. Groups and Permutations

## 10.1    Permutations and Groups

A permutation $f$ can be regarded as a function on the set $\{1, 2, \ldots, n\}$ such that $f(1), f(2), \ldots, f(n)$ is a reordering of $1, 2, \ldots, n$. We often represent the permutation $f$ by the following row and column notation:

$$\begin{array}{cccc} 1 & 2 & \ldots & n \\ f(1) & f(2) & \ldots & f(n) \end{array}.$$

For instance, the permutation $f(1) = 3$, $f(2) = 4$, $f(3) = 1$, $f(4) = 2$ is represented as

$$\begin{array}{cccc} 1 & 2 & 3 & 4 \\ 3 & 4 & 1 & 2 \end{array}.$$

Interchanging the orders of the columns in this representation does not change the permutation. For instance, the preceding permutation can also be written as

$$\begin{array}{cccc} 3 & 4 & 1 & 2 \\ 1 & 2 & 3 & 4 \end{array},$$

meaning that $f(3) = 1$, $f(4) = 2$, $f(1) = 3$, and $f(2) = 4$.

If $f$ and $g$ are permutations on the set $\{1, 2, \ldots, n\}$, then we define their composition $f \odot g$ to be the function such that

$$f \odot g(i) = f(g(i)).$$

Since $g(1), \ldots, g(n)$ is a rearrangement of $1, \ldots, n$ and $f$ is a permutation, it follows that $f(g(1)), \ldots, f(g(n))$ is also a rearrangement of $1, \ldots, n$. That is, $f \odot g$ is itself a permutation.

**Example 10.1a**    If $f$ is the permutation

$$\begin{array}{cccc} 1 & 2 & 3 & 4 \\ 3 & 4 & 1 & 2 \end{array}$$

and $g$ is the permutation

$$1\ 2\ 3\ 4$$
$$4\ 2\ 1\ 3$$

then $f \odot g$ is the permutation

$$1\ 2\ 3\ 4$$
$$2\ 4\ 3\ 1\,\dot{}$$

The preceding follows because

$$f \odot g(1) = f(g(1)) = f(4) = 2,$$
$$f \odot g(2) = f(g(2)) = f(2) = 4,$$

and so on.                                             □

The set of all $n!$ permutations on the set $\{1, 2, \ldots, n\}$ along with the composition operator is a "group," according to the following definition.

**Definition**    A nonempty set of elements $G$ along with a composition operator $\odot$ between elements of $G$ is said to be a *group* if the following conditions all hold.

(1) If $g \in G$ and $h \in G$, then $g \odot h \in G$.
(2) There is an element $I$ of $G$ (called the *identity element*) such that, for all $g \in G$,

$$g \odot I = I \odot g = g.$$

(3) For each element $g \in G$, there is an element $g^{-1} \in G$ (called the *inverse* of $g$) such that

$$g \odot g^{-1} = g^{-1} \odot g = I.$$

(4) For all elements $f, g, h$ in $G$, the *associative property*

$$(f \odot g) \odot h = f \odot (g \odot h)$$

holds.

**Example 10.1b**    (a) The set of all integers is a group when the composition operation is addition. The identity element of this group is the integer 0, and the inverse of the integer $n$ is $-n$.

(b) The set of all real numbers different from 0 forms a group when the composition operation is multiplication; 1 is the identity element, and $1/x$ is the inverse of $x$. □

To check that the set of permutations forms a group, note that if the permutation $I$ is defined by

$$I(i) = i, \quad i = 1, \ldots, n,$$

then $I$ satisfies the conditions of the identity element. Also, if $f$ is the permutation

$$\begin{array}{cccc} 1 & 2 & \ldots & n \\ f(1) & f(2) & \ldots & f(n) \end{array}$$

then the permutation defined by

$$\begin{array}{cccc} f(1) & f(2) & \ldots & f(n) \\ 1 & 2 & \ldots & n \end{array}$$

is the inverse permutation $f^{-1}$. In other words, $f^{-1}(i)$ is the value that $f$ maps into $i$; that is,

$$f(f^{-1}(i)) = f^{-1}(f(i)) = i.$$

**Example 10.1c** According to the preceding, if $f$ is the permutation

$$\begin{array}{cccc} 1 & 2 & 3 & 4 \\ 4 & 3 & 1 & 2 \end{array}$$

then $f^{-1}$ is the permutation

$$\begin{array}{cccc} 4 & 3 & 1 & 2 \\ 1 & 2 & 3 & 4 \end{array}.$$

As a check, note that

$$f \odot f^{-1}(1) = f(f^{-1}(1)) = f(3) = 1,$$
$$f^{-1} \odot f(1) = f^{-1}(f(1)) = f^{-1}(4) = 1;$$
$$f \odot f^{-1}(2) = f(f^{-1}(2)) = f(4) = 2,$$
$$f^{-1} \odot f(2) = f^{-1}(f(2)) = f^{-1}(3) = 2.$$

Similarly,

$$f \odot f^{-1}(3) = f^{-1} \odot f(3) = 3, \quad f \odot f^{-1}(4) = f^{-1} \odot f(4) = 4. \quad \square$$

To complete our verification that the set of all permutations constitutes a group with respect to the composition operation previously defined, we must check that condition (4) holds. To do so, let $f, g, h$ be permutations. Then

$$(f \odot g) \odot h(i) = (f \odot g)(h(i)) = f(g(h(i)))$$

and

$$f \odot (g \odot h)(i) = f(g \odot h(i)) = f(g(h(i))),$$

which verifies condition (4). Thus, the set of permutations on the set $\{1, 2, \ldots, n\}$ is a group; it is called the *symmetric group,* denoted $S_n$.

**Example 10.1d**    A group $G$ is said to be *abelian* if

$$f \odot g = g \odot f$$

for all $f, g \in G$. The group of integers with respect to the addition operation and the group of nonzero real numbers with respect to multiplication are both abelian groups, but the group of permutations is not abelian. For instance, for the permutations defined in Example 10.1a, we have that $g \odot f$ is the permutation

$$\begin{array}{cccc} 1 & 2 & 3 & 4 \\ 1 & 3 & 4 & 2 \end{array},$$

which, as seen from that example, is not equal to $f \odot g$.    $\square$

**Example 10.1e**    For a prime integer $p$, let $Z_p^* = \{1, 2, \ldots, p - 1\}$ and consider $Z_p^*$ along with the composition operator

$$i \odot j = ij \bmod p.$$

That is, $i \odot j$ is the remainder when the product of $i$ and $j$ is divided by $p$. For instance, if $p = 7$ then

$$5 \odot 6 = 30 \bmod 7 = 2.$$

Show that, when $p$ is prime, $Z_p^*$ is an abelian group.

**Solution.** We have $i \odot j = j \odot i$, so it suffices to show that $Z_p^*$ is a group. To do so, first note that if $i$ and $j$ are in $Z_p^*$ then so is $i \odot j$, provided that $ij \neq 0 \bmod p$. But if $ij = 0 \bmod p$ then $p$ divides $ij$, which implies (by the prime factorization theorem) that $p$ divides either $i$ or $j$. However, since both $i$ and $j$ are betweeen 1 and $p-1$, it is impossible for $p$ to divide either of them. Therefore we conclude that $i \odot j \in Z_p^*$. Clearly, 1 is the identity element.

In order to verify the associative property $(i \odot j) \odot k = i \odot (j \odot k)$, we will show that

$$(i \odot j) \odot k = ijk \bmod p.$$

To show this, use Euclid's division lemma (Proposition 1.4.1) to obtain that there are nonnegative integers $r_1$ and $r_2$, both less than $p$, such that

$$ij = q_1 p + r_1,$$
$$r_1 k = q_2 p + r_2.$$

Hence,

$$(i \odot j) \odot k = r_1 \odot k = r_2.$$

On the other hand,

$$ijk = (q_1 p + r_1)k$$
$$= q_1 pk + r_1 k$$
$$= q_1 pk + q_2 p + r_2$$
$$= (q_1 k + q_2)p + r_2,$$

thus showing that

$$(i \odot j) \odot k = r_2 = ijk \bmod p.$$

Since

$$i \odot (j \odot k) = (j \odot k) \odot i = jki \bmod p,$$

the associative property is verified.

It remains to show that every element in $Z_p^*$ has an inverse. To do so, fix $i$ ($1 \leq i < p$) and consider the function

$$f(j) = i \odot j, \quad j = 1, \ldots, p-1.$$

We claim that $f(1), \ldots, f(p-1)$ are all distinct. For suppose that, for some $1 \leq j < k < p$,

$$f(j) = f(k).$$

If so, then

$$ij = ik \bmod p,$$

which implies that

$$i(k - j) = 0 \bmod p.$$

That is, $p$ divides $i(k - j)$. Since $p$ is prime, this yields that either $p$ divides $i$ or $p$ divides $k - j$. But $i$ and $k - j$ are both less than $p$, so this is clearly impossible; thus, we can conclude that the $p-1$ values $f(1), \ldots, f(p-1)$ are all distinct. Each is one of the $p-1$ values $1, \ldots, p-1$, so it follows that each of these must be the value of exactly one of the $f(j)$. Thus there is a $j$, call it $j^*$, for which $f(j^*) = 1$. That is,

$$i \odot j^* = 1.$$

Because $i \odot j = j \odot i$, we can conclude that $j^* = i^{-1}$, thus completing the verification that $Z_p^*$ is a group.     □

If $g$ is an element of a group, let

$$g^1 = g$$

and, for $m > 1$, recursively define $g^m$ by

$$g^m = g \odot g^{m-1}.$$

In addition, define $g^{-2}$ by

$$g^{-2} = g^{-1} \odot g^{-1}$$

and, for $k > 2$,

$$g^{-k} = g^{-1} \odot g^{-(k-1)}.$$

It is not difficult to show (we leave it as an exercise) that

$$g^m \odot g^{-k} = g^{-k} \odot g^m = g^{m-k},$$

where

$$g^0 = I.$$

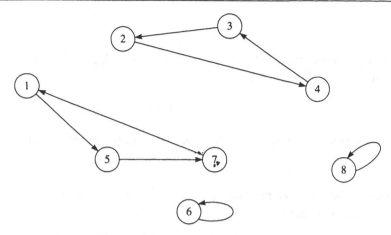

Figure 10.1: Digraph of a Permutation

## 10.2 Permutation Graphs

A permutation on $\{1, 2, \ldots, n\}$ can be represented by a directed graph with vertices $\{1, 2, \ldots, n\}$ and edges $(i, f(i))$, $i = 1, \ldots, n$. (Note that this directed graph allows edges from a vertex to itself.) Since $f$ is a permutation, it follows that there will be exactly one edge going into and one coming out of each vertex. For a given vertex $i$, consider the sequence of vertices $i, f(i), f^2(i), f^3(i), \ldots$ whose successive pairs are edges of the graph. Since there are only a finite number of vertices, there will eventually be a repeat vertex in this sequence; the first one to repeat must be vertex $i$, for otherwise there would be a vertex with at least two incoming edges. Thus, vertex $i$ (and every other vertex) will appear in the digraph as part of a cycle.

**Example 10.2a**   Consider the permutation

$$\begin{array}{cccccccc} 1 & 2 & 3 & 4 & 5 & 6 & 7 & 8 \\ 5 & 4 & 2 & 3 & 7 & 6 & 1 & 8 \end{array}.$$

Figure 10.1 gives the digraph representation of this permutation. Thus, every connected component of the permutation digraph is a cycle. As a result, we can also represent a permutation by giving its cycles. For instance, with the cycles denoted by the vertices within a bracket, the permutation of Example 10.2a can be represented as

$$[1 \ 5 \ 7] \ [2 \ 4 \ 3] \ [6] \ [8].$$

A useful convention is to list only the cycles of length greater than 1. We would thus use the notation

$$[1 \ 5 \ 7] \ [2 \ 4 \ 3]$$

for the preceding permutation. Also, we let [1] stand for the identity element *I*.    □

**Example 10.2b**    The permutation $f$ on the integers 1 through 10 given by the cycle notation

$$f = [1 \ 3 \ 6 \ 2] \ [5 \ 10]$$

is the permutation

$$
\begin{array}{cccccccccc}
1 & 2 & 3 & 4 & 5 & 6 & 7 & 8 & 9 & 10 \\
3 & 1 & 6 & 4 & 10 & 2 & 7 & 8 & 9 & 5
\end{array}
$$

Note that, if $i$ is not listed in the cycle notation for $f$, then $f(i) = i$.    □

A permutation having only one cycle of length greater than 1 is called a *cycle permutation*.

**Example 10.2c**    A cycle of length 2 is called a *transposition*. For instance, the permutation transposition $[i, j]$ is equivalently written as

$$f(i) = j, \quad f(j) = i, \quad f(k) = k,$$
$$k \neq i, \quad k \neq j.$$    □

## 10.3    Subgroups

If the set of elements $G$ is a group with respect to the operation $\odot$, and if $H$ is a subset of $G$ such that $H$ is also a group with respect to $\odot$, then we say that $H$ is a *subgroup* of $G$. The nonempty subset $H$ will be a subgroup provided that the following hold:

(a) if $h \in H$, then $h^{-1} \in H$;

(b) if $h \in H$ and $r \in H$, then $h \odot r \in H$.

To verify that (a) and (b) imply that $H$ is a subgroup, note that (a) and (b) together imply that $I \in H$.

Condition (b) by itself suffices when $G$ is a group with a finite number of elements.

**Proposition 10.3.1** *If $G$ is a finite group then $H$, a nonempty subset of $G$, is a subgroup if*

$$h \in H \ \& \ r \in H \implies h \odot r \in H.$$

***Proof.*** Let $h$ be an element of $H$. Then, by the condition of the proposition, it follows that $h, h^2, h^3, \ldots$ are all elements of $H$. Since $H$ is a subset of a finite set, it too is finite, so these elements can not all be distinct. Hence there must be values $k < m$ such that

$$h^k = h^m.$$

But this implies that

$$h^{-k} \cdot h^k = h^{-k} \cdot h^m$$

or (equivalently) that

$$I = h^{m-k},$$

implying that $I \in H$. In addition, since

$$I = h^{m-k} = h \odot h^{m-k-1},$$

it follows that

$$h^{-1} \odot I = h^{-1} \odot (h \odot h^{m-k-1}) = (h^{-1} \odot h) \odot h^{m-k-1} = I \odot h^{m-k-1}.$$

Therefore,

$$h^{-1} = h^{m-k-1},$$

which shows that $h^{-1} \in H$ and thus that $H$ is a group. $\square$

Permutation groups are particularly important because any finite group is isomorphic to a subgroup of a permutation group, where we say that two groups $G$ and $H$ are *isomorphic* if there is a one-to-one correspondence between their elements such that this correspondence is preserved under group compositions. In other words, for each $g \in G$, there is a matching element $h \in H$ (with the matching indicated by the notation $g \longleftrightarrow h$) such that every element in $H$ is matched to one and only one element in $G$, and if $g_1 \longleftrightarrow h_1$ and $g_2 \longleftrightarrow h_2$ then

$$g_1 \odot g_2 \longleftrightarrow h_1 \odot h_2.$$

Thus, the groups $G$ and $H$ are isomorphic if $H$ is, in essence, the same group as $G$ (except that the elements have been given different names). The concept of groups being isomorphic is important because it gives recognition to the fact that the same abstract group can appear in many different applications.

**Proposition 10.3.2** (Cayley's Group Isomorphic Theorem)    *Any finite group $G$ is isomorphic to a subgroup of a permutation group.*

**Proof.** If $G$ has $n$ elements, then arbitrarily rename these elements as $1, 2, \ldots, n$. Match element $j$ to the permutation $f_j$, where

$$f_j(i) = j \odot i.$$

To check that $f_j$ is a permutation, note that each of $f_j(1), \ldots, f_j(n)$ is one of the $n$ values $1, \ldots, n$. Thus, in order to show that $f_j$ is a permutation, we need only show that these values are all distinct. If $f_j(i) = f_j(k)$ then

$$j \odot i = j \odot k,$$

implying that

$$j^{-1} \odot j \odot i = j^{-1} \odot j \odot k$$

or

$$i = k.$$

Thus, we can conclude that $f_j$ is a permutation. It remains to show that the set of these permutations is a group and also that $f_{j \odot k} = f_j \cdot f_k$, where we have used the symbol $\cdot$ to indicate the composition function for permutations. However, this latter property implies that $f_j \cdot f_k$ is a member of the set of permutations $\{f_1, \ldots, f_n\}$; since this set is contained in $S_n$, by Proposition 10.3.1 we can conclude that $\{f_1, \ldots, f_n\}$ is a group. Therefore, to complete the proof we need only show that $f_{j\odot k} = f_j \cdot f_k$, which is equivalent to showing that, for each $i$ $(i = 1, \ldots, n)$,

$$f_{j\odot k}(i) = f_j \cdot f_k(i)$$

or (equivalently) that

$$(j \odot k) \odot i = f_j(k \odot i)$$

or

$$(j \odot k) \odot i = j \odot (k \odot i),$$

which follows from the associative property of groups. $\square$

The number of elements in a finite group is called the *order* of the group. It follows from the proof of Proposition 10.3.1 that, if $g$ is an element of a finite group $G$, then $\{g, g^2, g^3, \ldots\}$ will be a subgroup of $G$. If we define $r$ to be the smallest value of $i$ $(i \geq 1)$ such that

$$g^i = I,$$

then the values in the sequence $g, g^2, \ldots$ will repeat after the first $r$ values. As a result, we can conclude that the set $\{g, g^2, \ldots, g^r = I\}$ is a subgroup; it is called the *cyclic subgroup* generated by $g$. Since it can be shown that each of the values $g^1, g^2, \ldots, g^r$ are distinct, it follows that $r$ is equal to the number of elements of the subgroup. Since $r$ is the order of the subgroup generated by $g$, we often just say that $r$ is the order of the element $g$.

**Example 10.3a** Consider the set of permutations on the set $\{1, 2, 3\}$, and label its elements as follows:

$$I = 1\ 2\ 3,$$

$$a = 1\ 3\ 2,$$

$$b = 2\ 1\ 3,$$

$$c = 2\ 3\ 1,$$

$$d = 3\ 1\ 2,$$

$$e = 3\ 2\ 1.$$

(Note that we are identifying each permutation by its lower row when its upper row is 1 2 3.) The composition table for this group is given as follows, where the elements inside the table are the products $r \odot c$ ($r$ is the row and $c$ is the column element, so e.g. the table yields that $a \odot b = d$).

|   | $I$ | $a$ | $b$ | $c$ | $d$ | $e$ |
|---|---|---|---|---|---|---|
| $I$ | $I$ | $a$ | $b$ | $c$ | $d$ | $e$ |
| $a$ | $a$ | $I$ | $d$ | $e$ | $b$ | $c$ |
| $b$ | $b$ | $c$ | $I$ | $a$ | $e$ | $d$ |
| $c$ | $c$ | $b$ | $e$ | $d$ | $I$ | $a$ |
| $d$ | $d$ | $e$ | $a$ | $I$ | $c$ | $b$ |
| $e$ | $e$ | $d$ | $c$ | $b$ | $a$ | $I$ |

From the preceding table, we see that $d^2 = c$ and $d^3 = d \odot d^2 = d \odot c = I$, thus showing that the cyclic subgroup generated by $d$ is $\{d, c, I\}$. Similarly, we see that the cyclic subgroup generated by

$I$ is $\{I\}$,
$a$ is $\{a, I\}$,
$b$ is $\{b, I\}$,
$c$ is $\{c, d, I\}$,
$d$ is $\{d, c, I\}$,
$e$ is $\{e, I\}$.    □

The following proposition is useful.

**Proposition 10.3.3**    *Let r be the order of the group element g. If $g^m = I$ then r is a divisor of m.*

**Proof.**  By Euclid's division lemma, we can write

$$m = qr + s,$$

where $0 \le s < r$. Hence,

$$I = g^{qr+s} = g^{qr} \odot g^s = (g^r)^q \odot g^s = I^q \odot g^s = I \odot g^s = g^s.$$

Since $r$ is the smallest positive integer such that $q^r = I$, the preceding implies that $s = 0$ and thus that $r$ divides $m$.    □

Let us now consider how to determine the order of a permutation. To begin, consider a cycle permutation $f = [i_1, i_2, \ldots, i_k]$. Then, ignoring those elements that are unchanged under $f$, we can write

$$f^0 = \begin{array}{ccccccc} i_1 & i_2 & i_3 & \cdots & i_{k-2} & i_{k-1} & i_k \\ i_1 & i_2 & i_3 & \cdots & i_{k-2} & i_{k-1} & i_k \end{array},$$

$$f^1 = \begin{array}{ccccccc} i_1 & i_2 & i_3 & \cdots & i_{k-2} & i_{k-1} & i_k \\ i_2 & i_3 & i_4 & \cdots & i_{k-1} & i_k & i_1 \end{array},$$

$$f^2 = \begin{array}{ccccccc} i_1 & i_2 & i_3 & \cdots & i_{k-2} & i_{k-1} & i_k \\ i_3 & i_4 & i_5 & \cdots & i_k & i_1 & i_2 \end{array},$$

$$f^3 = \begin{array}{ccccccc} i_1 & i_2 & i_3 & \cdots & i_{k-2} & i_{k-1} & i_k \\ i_4 & i_5 & i_6 & \cdots & i_1 & i_2 & i_3 \end{array},$$

and so on. It is easy to see from this progression that the first positive $m$ for which $f^m = I$ is $m = k$. Thus, the order of a cycle permutation is the length of the cycle. More generally, we have the following result.

**Proposition 10.3.4** *The order of a permutation is the least common multiple of the lengths of its disjoint cycles.*

**Proof.** Suppose that $f$ is the composition of $s$ disjoint cycles $f_1, f_2, \ldots, f_s$ of respective lengths $n_1, n_2, \ldots, n_s$. Then it is easy to see that

$$f^m = f_1^m \odot f_2^m \odot \cdots \odot f_s^m.$$

The order of $f_i$ is $n_i$, so it follows that $f_i^m$ will be the identity permutation if and only if $m$ is a multiple of $n_i$. (The "only if" part follows from Proposition 10.3.3.) Therefore, $f^m = I$ if and only if $m$ is a multiple of all the cycle lengths, implying that the smallest such $m$ is the least common multiple. $\qquad\square$

## 10.4 Lagrange's Theorem

Let $H$ be a subgroup of the group $G$. For an element $g \in G$, define the set $g \odot H$ by

$$g \odot H = \{g \odot h : h \in H\}.$$

The set $g \odot H$ is called a *left coset* of the group $G$. We will now prove two lemmas regarding these sets.

**Lemma 10.4.1**  *If $H$ is a subgroup of the finite group $G$ then, for all $g \in G$,*

$$|g \odot H| = |H|,$$

*where $|S|$ is equal to the number of elements in the set $S$.*

***Proof.*** If $g \odot h_1 = g \odot h_2$, then

$$g^{-1} \odot (g \odot h_1) = g^{-1} \odot (g \odot h_2)$$

or (equivalently)

$$(g^{-1} \odot g) \odot h_1 = (g^{-1} \odot g) \odot h_2$$

or

$$h_1 = h_2.$$

Therefore, all of the elements $g \odot h$ ($h \in H$) are distinct, and the result follows.    □

**Lemma 10.4.2**  *If $H$ is a subgroup of the finite group $G$ then $g_1 \odot H$ and $g_2 \odot H$ are either disjoint or identical.*

***Proof.*** Suppose these sets have an element in common, say $x$. Then, for some $h_1 \in H$ and $h_2 \in H$,

$$x = g_1 \odot h_1 = g_2 \odot h_2,$$

which implies, upon composition with $h_1^{-1}$, that

$$g_1 = g_2 \odot h_2 \odot h_1^{-1}.$$

Hence, for any $h \in H$,

$$g_1 \odot h = g_2 \odot h_2 \odot h_1^{-1} \odot h.$$

Since $H$ is a group, $h_2 \odot h_1^{-1} \odot h \in H$, which shows that

$$g_1 \odot H \subset g_2 \odot H.$$

By reversing the roles of $g_1$ and $g_2$, we likewise obtain that

$$g_2 \odot H \subset g_1 \odot H.$$

Thus,

$$g_1 \odot H = g_2 \odot H$$

and the result is proven.    □

The preceding lemmas imply the important group theoretic result known as Lagrange's theorem.

**Theorem 10.4.1** (Lagrange's Theorem)   *If $H$ is a subgroup of the finite group $G$, then $|H|$ divides $|G|$. That is, the order of a subgroup divides the order of the group.*

*Proof.* Let $k$ equal the number of distinct sets of the form $g \odot H$. As these sets each contain $|H|$ elements (by Lemma 10.4.1), are mutually exclusive (by Lemma 10.4.2), and together contain all of the elements of $G$ (since $g \in g \odot H$), it follows that $|G| = k|H|$.    □

Lagrange's theorem has some interesting consequences.

**Corollary 10.4.1**   *If $G$ is a finite group of order $n$ then, for any $g \in G$,*

$$g^n = I.$$

*Proof.* By Lagrange's theorem, the order of the cyclic subgroup

$$g, g^2, \ldots, g^r = I$$

divides $n$. Therefore, $n = rk$ for some $k$ and so

$$g^n = (g^r)^k = I^k = I.$$    □

**Corollary 10.4.2**   *Let $G$ be a finite group of order $n$. If $n$ is prime, then $G$ is cyclic and thus abelian.*

*Proof.* If $G$ consists only of the identity element, then the result is true. If not, let $g$ ($g \neq I$) be an element of $G$ and consider the cyclic subgroup of distinct elements $S = \{g, g^2, \ldots, g^r = I\}$. By Lagrange's theorem, $r$ divides $n$, which implies (since $n$ is prime) that $n = r$. Thus, $G = S$ and the result is proved.                                                   □

**Example 10.4a**    In Example 2.2f we gave a combinatorial argument to prove Fermat's little theorem, which states that if $p$ is prime and $n$ is not a multiple of $p$ then $n^{p-1} - 1$ is divisible by $p$. We will now present a second argument that uses the group of Example 10.1e along with Lagrange's theorem. In Example 10.1e, we showed that if $p$ is prime then the set $\{1, 2, \ldots, p - 1\}$ is a group with respect to

$$i \odot j = ij \bmod p.$$

The order of this group is $p - 1$ and the identity element is 1, so it follows from Corollary 10.4.1 that

$$n \odot n \odot \cdots \odot n = 1,$$

where $n$ appears $p - 1$ times on the LHS. Since this is equivalent to

$$n^{p-1} = 1 \bmod p,$$

we see that $n^{p-1} - 1$ is divisible by $p$ for any integer $n$ in the group. This gives us Fermat's little theorem only when $1 \le n < p$, but we could have let the group consist of the integers $kp + 1, kp + 2, \ldots, kp + p - 1$, with

$$i \odot j = (ij \bmod p) + kp.$$

The identity of this group would be $kp + 1$, and Corollary 10.4.1 gives that, for any $n$ in the group,

$$n^{p-1} = kp + 1 \bmod p.$$

As $kp + 1 = 1 \bmod p$, Fermat's little theorem is obtained.            □

**Example 10.4b** *Fermat Numbers*    For some nonnegative integer $m$, let $p = 2^m + 1$ and suppose that $p$ is prime. Now consider the group $\{1, \ldots, p - 1\}$ whose composition operation is multiplication modulo $p$, and let $o(2)$ be the order of 2 as a member of this group.

*Claim:* $o(2) = 2m$. To verify the claim, note first that

$$2^{2m} = 1 \bmod p$$

is equivalent to the statement that

$$2^{2m} - 1 \text{ is divisible by } p,$$

and this follows since $p = 2^m + 1$ and

$$2^{2m} - 1 = (2^m - 1)(2^m + 1).$$

Thus, with respect to the group under consideration,

$$2^{2m} = I;$$

by Proposition 10.3.3, this implies that $o(2)$ is a divisor of $2m$. But $o(2) > m$, since

$$2^k \neq 1 \bmod p \quad \text{for } k = 1, \ldots, m.$$

Hence, we can conclude that $o(2) = 2m$. By Lagrange's theorem, this implies that $2m$ is a divisor of $p - 1 = 2^m$, yielding (by the prime factorization theorem) that $m$ must be a multiple of 2 – say, $m = 2^n$.

Thus we have shown that the only prime numbers of the form $2^m + 1$ are of the type $2^{2^n} + 1$. Numbers of this type are called *Fermat numbers*. Fermat – noting that

$$2^{2^0} + 1 = 2 + 1 = 3,$$

$$2^{2^1} + 1 = 2^2 + 1 = 5,$$

$$2^{2^2} + 1 = 2^4 + 1 = 17,$$

$$2^{2^3} + 1 = 2^8 + 1 = 257,$$

$$2^{2^4} + 1 = 2^{16} + 1 = 65,537$$

are all prime – conjectured the possibility that all Fermat numbers are prime. Euler, however, refuted that possibility by showing that

$$2^{2^5} + 1 = 2^{32} + 1 = 4,294,967,297$$

254     *Groups and Permutations*

is divisible by 641. Indeed, no Fermat number having $n > 4$ has ever been shown to be prime. (It is presently unknown whether the Fermat number with $n = 20$ is prime.) $\qquad\qquad\qquad\qquad\square$

*Normal* subgroups have a particular importance in group theory. They are defined as follows.

**Definition**  A subgroup $N$ of the group $G$ is said to be a *normal* subgroup if, for all $g \in G$,

$$g \odot N = N \odot g.$$

That is, $N$ is normal if, for all $g \in G$,

$$\{g \odot h : h \in N\} = \{h \odot g : h \in N\}.$$

## 10.5    The Alternating Subgroup

In this section we define the alternating subgroup of the symmetric permutation group $S_n$. However, before doing so, we need the concepts of even and odd permutations. We start by defining the polynomial function $P(x_1, \ldots, x_n)$ as

$$P(x_1, \ldots, x_n) = \prod_{i<j}(x_i - x_j),$$

where the product is defined over all the $\binom{n}{2}$ pairs $i, j$ for which $i < j$. For instance,

$$P(x_1, x_2, x_3) = (x_1 - x_2)(x_1 - x_3)(x_2 - x_3).$$

For any permutation $f \in S_n$, let

$$P_f(x_1, \ldots, x_n) = P(x_{f(1)}, \ldots, x_{f(n)}) = \prod_{i<j}(x_{f(i)} - x_{f(j)}).$$

Call $f$ *even* if
$$P_f(x_1, \ldots, x_n) = P(x_1, \ldots, x_n)$$

and call $f$ *odd* otherwise. For each distinct pair $i \neq j$, either $(x_i - x_j)$ or its negative will be one of the $\binom{n}{2}$ products in both $P(x_1, \ldots, x_n)$ and $P_f(x_1, \ldots, x_n)$, so it follows that

$$P_f(x_1, \ldots, x_n) = -P(x_1, \ldots, x_n)$$

when $f$ is an odd permutation.

Let $S(f)$, the *parity sign* of the permutation $f$, equal 1 if $f$ is even or $-1$ if $f$ is odd. Thus,

$$P_f(x_1, \ldots, x_n) = S(f)P(x_1, \ldots, x_n).$$

Now, for permutations $f$ and $g$,

$$
\begin{aligned}
P_{f \odot g}(x_1, \ldots, x_n) &= \prod_{i<j}(x_{f \odot g(i)} - x_{f \odot g(j)}) \\
&= P_f(x_{g(1)}, \ldots, x_{g(n)}) \\
&= S(f)P(x_{g(1)}, \ldots, x_{g(n)}) \\
&= S(f)S(g)P(x_1, \ldots, x_n).
\end{aligned}
$$

Consequently, we see that

$$S(f \odot g) = S(f)S(g). \tag{10.1}$$

Thus, the composition of two permutations is even when the permutations are both even or both odd, and the composition is odd when the permutations are of opposite parity sign.

**Lemma 10.5.1**  *For any permutation $g$,*

$$S(g) = S(g^{-1}).$$

***Proof.*** Since $I = g \odot g^{-1}$, it follows that

$$S(I) = S(g)S(g^{-1}).$$

As $S(I) = 1$, the result follows.  $\square$

**Proposition 10.5.1**  *The set of even permutations, designated as $A_n$, is a normal subgroup of $S_n$. It is called the alternating subgroup.*

***Proof.*** Equation (10.1) along with Proposition 10.3.1 shows that $A_n$ is a subgroup of $S_n$. To show that it is a normal subgroup, we need to show that $g \odot A_n = A_n \odot g$ for every permutation $g$. Let $f \in A_n$, and note that

$$g \odot f = x \odot g$$

has the solution

$$x = g \odot f \odot g^{-1}.$$

To show that $x \in A_n$, note that

$$S(x) = S(g)S(f \odot g^{-1})$$
$$= S(g)S(f)S(g^{-1})$$
$$= S(f) \quad \text{(by Lemma 10.5.1)}$$
$$= 1 \quad \text{(since } f \in A_n \text{)}.$$

Hence $x \in A_n$, showing that $g \odot A_n \subset A_n \odot g$. A similar argument gives that $A_n \odot g \subset g \odot A_n$, and the result follows.  □

Let $f$ be a transposition permutation, say $f = [k, m]$, where $k < m$. For this permutation, $P_f$ is obtainable from $P$ upon replacing $x_k$ by $x_m$ and $x_m$ by $x_k$ wherever they appear in $P$. Thus, any factor of $P$ of the form $x_i - x_j$ (where neither $i$ nor $j$ is $k$ or $m$) will remain unchanged. When $i < k$, the product $(x_i - x_k)(x_i - x_m)$ will remain unchanged; when $k < i < m$, the product $(x_k - x_i)(x_i - x_m)$ will also remain unchanged (as it will be changed to $(x_m - x_i)(x_i - x_k)$). Thus, aside from the factor $x_k - x_m$ (which is changed into its negative), the product of the other factors of $P$ remains unchanged. Hence, if $f$ is a transposition permutation then $P_f = -P$. That is, all transposition permutations are odd.

It follows from the preceding that, if $f$ is an even permutation, then the permutation

$$\begin{matrix} 1 & 2 & 3 & \dots & n \\ f(2) & f(1) & f(3) & \dots & f(n) \end{matrix}$$

is odd (i.e., the permutation $[f(1), f(2)] \odot f$ is odd). This implies that there are the same number of odd and even permutations. Since there are a total of $n!$ permutations, we obtain the following.

**Proposition 10.5.2**  *There are $n!/2$ even permutations and the same number of odd ones.*

Any cycle permutation can be written as a composition of transposition permutations. For instance,

$$[1, 2, 3] = [1, 3][1, 2]$$

or (equivalently)

$$1\ 2\ 3 = 1\ 2\ 3 \odot 1\ 2\ 3,$$

$$2\ 3\ 1 = 3\ 2\ 1 \odot 2\ 1\ 3.$$

In general, we can write any cycle permutation $[i_1, i_2, \ldots, i_k]$ in the following manner:

$$[i_1, i_2, \ldots, i_k] = [i_1, i_k] \odot [i_1, i_{k-1}] \odot \cdots \odot [i_1, i_3] \odot [i_1, i_2].$$

Therefore, any cycle of length $k$ can be expressed as the product of $k - 1$ transpositions. As each transposition permutation changes the sign of $P$, it follows that the sign of $P$ remains unchanged if and only if $k - 1$ is even. That is, the sign of $P$ remains unchanged if and only if the cycle length is odd. As a result, since every permutation can be expressed as a product of disjoint cycles, we have shown the following.

**Proposition 10.5.3** *A permutation is even if and only if, in its expression as the composition of disjoint cycles, there are an even number of cycles of even length.*

Another way of determining the parity (even or odd) of a permutation is to count the number of different inversions it introduces. For a permutation $f$, consider the sequence

$$f(1)\ f(2)\ \ldots\ f(n).$$

We say that there are $k$ inversions introduced by $i$ if exactly $k$ of the values that precede $i$ in the foregoing sequence are greater than $i$. As an example, for the permutation

$$1\ 2\ 3\ 4\ 5\ 6$$
$$4\ 2\ 1\ 6\ 5\ 3$$

we have

- 2 inversions introduced by 1,
- 1 inversion introduced by 2,
- 3 inversions introduced by 3,

- 0 inversions introduced by 4,
- 1 inversion introduced by 5, and
- 0 inversions introduced by 6.

The total number of inversions is $2 + 1 + 3 + 0 + 1 + 0 = 7$. In general, let $V_f(i)$ denote the number of inversions introduced by $i$ in the permutation $f$, and let

$$V_f = \sum_{i=1}^{n} V_f(i)$$

be the total number of inversions in $f$.

Now consider what happens when we take the composition of the permutation $f$ and the transposition permutation $[i, j]$. For instance, consider

$$
\begin{matrix} 1 & 2 & 3 & 4 & 5 & 6 \\ 4 & 2 & 1 & 6 & 5 & 3 \end{matrix} \odot [2, 4] =
\begin{matrix} 1 & 2 & 3 & 4 & 5 & 6 \\ 4 & 6 & 1 & 2 & 5 & 3 \end{matrix}.
$$

That is, $f \odot [i, j]$ interchanges the values of $f(i)$ and $f(j)$. Thus, there is a composition of $f$ and $V_f(1)$ transpositions that gives a permutation whose first element is 1. For instance, identifying the permutation by the vector $f(i)$, $i = 1, \ldots, n$, we have

$$4\ 2\ 1\ 3 \odot [2, 3] \odot [1, 2] = 1\ 4\ 2\ 3.$$

An additional $V_f(2)$ compositions with transpositions can be chosen to obtain a permutation whose first two elements are $1, 2$. Continuing in this fashion shows that transpositions can be chosen so that, after $V_f$ compositions with these transpositions, we obtain the identity permutation $I$. But $I$ is even (and so has parity sign 1) whereas each transposition is odd, implying that

$$1 = S(I) = (-1)^{V_f} S(f).$$

Hence, we have shown the following.

**Proposition 10.5.4**   *The permutation $f$ is even if it has an even number of inversions and is odd if it has an odd number of inversions.*

The following result can be proven.

**Theorem 10.5.1** (Galois's Theorem) *If $n > 4$, then the only normal subgroups of $A_n$ are $A_n$ and $\{I\}$.*

Galois's theorem is a very important result in the history of algebra, for it played a key part in Galois's proof that polynomial equations of degree 5 or larger are not generally solvable in terms of radicals. That is, when $n \geq 5$, equations of the form

$$a_0 + a_1 x + a_2 x^2 + \cdots + a_n x^n = 0$$

cannot generally be solved by a sequence of rational operations (addition, subtraction, multiplication, and division) and the taking of $n$th roots of quantities already known. However, for $n \leq 4$ such equations can be solved in terms of radicals; for instance, we have the well-known solution

$$x = \frac{-a_1 \pm \sqrt{a_1^2 - 4a_0 a_2}}{2a_0}$$

when $n = 2$.

## 10.6    Exercises

Unless otherwise noted, assume that all elements are members of a group.

**Exercise 10.1**    Show that, for elements $r$ and $s$,

(a) $g^{r+s} = g^r \odot g^s$;
(b) $(g^r)^s = g^{rs}$.

**Exercise 10.2**    Show that

$$(f \odot g)^{-1} = g^{-1} \odot f^{-1}.$$

**Exercise 10.3**    Construct an example to show that, in general,

$$(f \odot g)^r \neq f^r \odot g^r.$$

Give a condition under which the preceding is true, and prove the result.

**Exercise 10.4**   If $r$ is the smallest value of $i$ ($i \geq 1$) such that $g^i = I$, show that $g, g^2, \ldots, g^r$ are distinct.

**Exercise 10.5**   Let $Z_n = \{0, 1, \ldots, n-1\}$, and show that $Z_n$ is a group with respect to the composition

$$x \odot y = \begin{cases} x + y & \text{if } x + y < n, \\ x + y - n & \text{if } x + y \geq n. \end{cases}$$

**Exercise 10.6**   Give the multiplication table for the group $Z_4$.

**Exercise 10.7**   Verify that the set $\{1, 3, 7, 9\}$ is a group under multiplication modulo 20.

**Exercise 10.8**   Determine which of the following sets are groups under multiplication modulo 14:

$$\{1, 3, 5\}, \quad \{1, 3, 5, 7\}, \quad \{1, 7, 13\}, \quad \{1, 9, 11, 13\}.$$

**Exercise 10.9**   Give the multiplication table for $S_3$.

**Exercise 10.10**   Show that all permutations in $S_3$ are cycles. Is the same true for those in $S_4$?

**Exercise 10.11**   Draw the digraph for the permutation

$$\begin{array}{cccccccc} 1 & 2 & 3 & 4 & 5 & 6 & 7 & 8 \\ 4 & 2 & 1 & 6 & 8 & 3 & 5 & 7 \end{array}.$$

**Exercise 10.12**   Show that, for $f$ and $g$ in $S_n$, $f \odot g = g \odot f$ when $g$ is a transposition.

**Exercise 10.13**   Find the orders of the elements 0, 1, 2, 3 of the group $Z_4$ defined in Exercise 10.5.

**Exercise 10.14**   Show that, if $H$ is a nonempty subset of the group $G$, then $H$ is a subgroup if $a \in H$ and $b \in H$ imply that $a \odot b^{-1} \in H$.

**Exercise 10.15**   Find all subgroups of $S_3$.

**Exercise 10.16**   The *center* of a group $G$ is defined to consist of all elements $g \in G$ such that

$$g \odot h = h \odot g \quad \text{for all } h \in G.$$

Show that the center of a group is a subgroup.

**Exercise 10.17**   If the groups $G$ and $H$ are isomorphic, show that their identity elements are matched. Then show that, if $g \longleftrightarrow h$, we also have $g^{-1} \longleftrightarrow h^{-1}$.

**Exercise 10.18**   Show that the group of integers $\{0, 1, 2, 3\}$ with group composition being addition mod 4 is isomorphic to the group of integers $\{1, 2, 3, 4\}$ with group composition being multiplication mod 4.
   *Hint:* Start by writing down the composition tables for these groups.

**Exercise 10.19**   Show that the group of positive real numbers with multiplication as group composition is isomorphic to the group of all real numbers with addition as group composition.

**Exercise 10.20**   Let $G$ be a cyclic group of order $n$. If $d$ is a positive divisor of $n$, show that $G$ has a subgroup of order $d$.

**Exercise 10.21**   Show that the order of the group element $g$ is equal to the order of $g^{-1}$.

**Exercise 10.22**   Show that a group of order $p^k$ has a subgroup of order $p$ when $p$ is prime.

**Exercise 10.23**   Show that, in a group of order $2n$, there is an element besides the identity that is its own inverse.

**Exercise 10.24**   Show that $\{I\}$ and $G$ are both normal subgroups of the group $G$.

**Exercise 10.25**   Show that the center of a group is a normal subgroup.

**Exercise 10.26**   If $N$ is a normal subgroup of $G$ and if $H$ is any other subgroup, show that $N \odot H$ is a subgroup.

**Exercise 10.27**   Show, in Exercise 10.26, that if $H$ is also normal then so is $N \odot H$.

**Exercise 10.28**   Let $N$ be a normal subgroup of $G$. Show that, for $f$ and $g$ in $G$,

$$(f \odot N) \odot (g \odot N) = (f \odot g) \odot N.$$

**Exercise 10.29**   Let $Z$ be the group consisting of all the integers, with $\odot$ representing addition.

(a)  Describe in words the set $n \odot Z$.

(b)  Show that $n \odot Z$ is a normal subgroup.

**Exercise 10.30**   Show that $N$ is a normal subgroup of $G$ if and only if (i) it is a subgroup of $G$ and (ii) for all $g \in G$,

$$H = g \odot H \odot g^{-1},$$

where $g \odot H \odot g^{-1} = \{g \odot h \odot g^{-1}, h \in H\}$.

**Exercise 10.31**   Find the parity (even or odd) of the permutation

$$\begin{array}{cccccccccc} 1 & 2 & 3 & 4 & 5 & 6 & 7 & 8 & 9 & 10 \\ 4 & 9 & 7 & 6 & 8 & 3 & 1 & 5 & 10 & 2 \end{array}$$

(a)  by using its representation as the composition of unique cycles;

(b)  by counting its number of inversions.

# Index